introduction to applied numerical analysis

mcgraw-hill computer science series

RICHARD W. HAMMING
Bell Telephone Laboratories

EDWARD A. FEIGENBAUM
Stanford University

introduction to applied numerical analysis

richard w. hamming

Bell Telephone Laboratories, Incorporated
and
City College of New York

mcgraw-hill book company

New York, St. Louis, San Francisco, Düsseldorf, Johannesburg,
Kuala Lumpur, London, Mexico, Montreal, New Delhi, Panama,
Rio de Janeiro, Singapore, Sydney, Toronto

introduction to applied numerical analysis

Library of Congress Catalog Card Number 78-127970
07-025889-9

2 3 4 5 6 7 8 9 0 M A M M 7 9 8 7 6 5 4 3 2 1

This book was set in Helvetica by Textbook Services, Inc., and printed and bound by
The Maple Press Company. The designer was Paula Tuerk; the drawings were done by B. Handelman Associates, Inc. The editors were Richard F. Dojny and Joan DeMattia. Matt Martino supervised production.

preface

The appearance of yet another "Introduction to Applied Numerical Analysis" (even if the word "Applied" distinguishes it from most of the other books) requires justification. The justifications can be grouped under the following headings:

1 *The Subject Matter Covered* The material in this book was selected by considering those topics which the practicing scientist or engineer seems to need most, and which at the same time can be reasonably treated within the limits of a one-term course. In the belief that for many students this may well be the terminal course, many topics are treated rather lightly, instead of a few topics very intensively. The interrelationship of the material and its treatment has further affected the selection. The result is a somewhat unconventional set of topics for a first course.

2 *The Treatment of the Material* There is a strong effort to treat the material as simply as possible and in a uniform way, rather than as a set of unrelated topics. Also, some effort is made to prepare the student to face new problems that are not treated in the book.

3 *The Use of Illustrations* There is a deliberate attempt to teach as much as possible through the use of pictures and simple numerical illustrations, rather than through the more usual use of equations and words. It is hoped that this method will make the learning process easier as well as longer lasting.

4 *The Viewpoint Adopted* The viewpoint adopted throughout the text is that applied numerical analysis is a subject in itself and is neither a part of programming nor a branch of mathematics. These differences are deliberately stressed since most students seem to feel that applied numerical analysis is closely related to mathematics or programming, or both, and they need to be warned repeatedly that in practice there are many significant differences.

5 *The Separation of Programming from Numerical Analysis* Teaching programming along with numerical analysis, as is so often tried today, violates a well-known pedagogical principle

that you should not try to teach more than one new idea at a time. Such attempts usually introduce a new programming idea and then treat the numerical illustration in a superficial and often false way. As a result, the student emerges with a wrong conception of what numerical analysis is and how to use it practically.

The material in this text has been covered a number of times in fourteen 2-hour lectures to an evening class of advanced undergraduate and graduate engineers. Thus, in the usual 3-hour course there may be time for an additional 14 lectures on the local high-level language, Fortran, Algol, Mad, PL/1, or Basic, along with the essential parts of the local monitor system. Probably, these lectures on programming should be given by an instructor other than the one who gives the numerical analysis lectures, because when introducing the beginner to a computer, the teacher is justly interested in showing off the power and the range of application that the computer has, and therefore he usually does not have a real feeling for the details that occur in applied numerical analysis (which uses a small fraction of the flexibility of the machine). This plan of teaching programming and numerical analysis separately but within a single 3-hour course can probably be done with well-prepared, more advanced students; with not well-prepared, less advanced students, the text could be used to fill the entire course and the programming could be taught separately.

Few people today are an island unto themselves, and of necessity much of what they know and believe is learned from their colleagues. Thus, I am especially indebted to my colleagues at the Bell Telephone Laboratories and to the environment that encourages independent thinking. Many friends and students have made comments on earlier drafts, but as usual they are not responsible for the faults. In particular, I also wish to thank Prof. Roger Pinkham of Stevens Institute for helping me at every stage of the work, and for encouraging me to continue to the end. Others to whom I owe particular thanks are John Athanassopoulos for most of Chap. 9; M. P. Epstein, J. F. Kaiser, R. G. Kayel, S. P. Morgan, and D. G. Schweikert for comments on the manuscript; and Mrs. Dorothy Luciani and Miss Geraldine Marky for first-class secretarial help.

richard w. hamming

contents

introduction to applied numerical analysis

ROUNDOFF AND FUNCTION EVALUATION

1.1 The computer revolution

The impression an average scientist or engineer has when he first meets a large-scale, high-speed digital computer is that it is a machine for doing arithmetic, rapidly and accurately to be sure, but merely the same kind of arithmetic that he has done in the past by hand. In some respects this impression is correct, but in many respects it completely misses the significance of the computer revolution. The reason it is correct is that the machine must be given detailed instructions on exactly what to do to which numbers, often in far more detail than you would give to an assistant if you were directing his work. The reason it is false is that it fails to take into consideration the remarkable effects that arise from the tremendous increase in the speed and accuracy by machine over hand calculation.

TYPICAL RATES FOR MULTIPLYING

How	Rate
Hand calculation	1 per 20 seconds
Relay machines	1 per second
Electronic computers (1969)	10^6 (1 million) per second

Large changes are often measured in units of an "order of magnitude change," one order of magnitude meaning a factor of 10, two orders of magnitude a factor of 100, and so forth. Modern computers are more than a million times faster than hand calculation, six orders of magnitude faster.

UNITS OF TIME		TYPICAL RATES OF TRAVEL	
Millisecond	$\frac{1}{1000} = 10^{-3}$	Walking	4 mph
Microsecond	10^{-6}	Automobile	40 mph
Nanosecond	10^{-9}	Airplane	400 mph
Picosecond	10^{-12}		

A million arithmetic operations in a second is not unusual, or to put it another way, an arithmetic operation per microsecond. To grasp the meaning of a microsecond, consider the fact that there are more microseconds in an hour (3.6×10^9) than there are seconds in a century (3.16×10^9).

It is customarily recognized that a change of a single order of magnitude produces fundamentally new effects in a field. Thus, the change of more than six orders of magnitude in computer speed over hand-calculating speed has brought a great many new effects. Again to be dramatic, in 1 second a machine can do more arithmetic (and without making mistakes) than the average student will do by hand in his entire lifetime.

Machine computation costs per arithmetic operation are typically $\frac{1}{10,000} =$ 10^{-4} *that of hand calculation—a change of four orders of magnitude!*

It is not the purpose of this book to dwell on the glamorous aspects of computing; rather, the purpose is to provide an approach to machine computation that does not, as so often happens, either gloss over the small but significant details of computing with a modern digital computer or get trapped in the details of a particular machine and monitor system. The book tries to prepare the student to use machines not only for today's practical problems but also for tomorrow's for which he will need both specific rules and some understanding of the fundamentals of applied numerical analysis.

1.2   Mathematics versus numerical analysis

Mathematics (especially as it is currently taught) and numerical analysis differ from each other more than is usually realized. The most obvious differences are that mathematics regularly uses the infinite, both for the representation of numbers and for processes, whereas computing is necessarily done on a finite machine in a finite time. The finite representation of numbers in the machine leads to **roundoff**

errors, whereas the finite representation of processes leads to **truncation errors**.

In mathematics:

Typical numbers

$$\pi = 3.14159265358979323846264\cdots$$

$$\tfrac{1}{3} = 0.333333 \cdots$$

Typical processes

$$\frac{dy}{dx} = \lim_{\Delta x \to 0} \left\{ \frac{\Delta y}{\Delta x} \right\}$$

$$\int_a^b f(x)\ dx = \lim_{|\Delta x_i| \to 0} \sum_{i=1}^{N} f(x_i)\ \Delta x_i$$

This small roundoff effect produces an all-pervading "noise" in numerical analysis. Much of mathematics is very sensitive to small changes; it is vulnerable to noise or, as the statisticians say, "It is not robust."

In mathematics if

$$A = B$$
$$B = C$$
$$\cdot$$
$$\cdot$$
$$\cdot$$
$$Y = Z$$

then

$$A = Z$$

but in the presence of "noise" this is not necessarily true.

This difference is **fundamental**, and the failure to recognize that essentially only the noise-resistant parts of mathematics are relevant to applied numerical analysis (as well as to almost any application of mathematics to the real world) causes much confusion.

Other differences arise from different meanings for the same words in the two fields. As we shall show in Sec. 1.10, the expression "the zeros of a polynomial" may have a number of different meanings in computing, whereas it has a single meaning in mathematics. Again, as we shall see in Chaps. 4 and 13, the fundamental theorem of algebra (a polynomial of degree n has exactly n zeros) has a tendency to break down in the presence of noise when the degree of the polynomial becomes fairly large.

Sometimes the mathematical object is like a needle in a haystack when you actually try to find it.

Still other differences between mathematics and computing are in their basic aims and goals and in matters of taste and elegance as well.

Existence theorems are usually not sufficient in computing.

Mathematics prefers exact, precise theorems; applied numerical analysis must at times use plausible, heuristic methods.

The consequences of all these differences are at times quite serious and can lead to grave misunderstandings. Sometimes computations give results close to the mathematician's ideal system, but sometimes the results are far apart. When they differ, it may not be the computing that is wrong; rather, the mathematician's model, exact though it is to him, may be the sole source of error! Thus, neither computing nor mathematics is a safe crutch to lean upon to avoid careful thinking.

The first part of this chapter is concerned with the machine itself and how roundoff and related effects can influence the kinds of answers we get. The second part of the chapter is concerned with how to use the machine for the very simple process of evaluating a function. Function evaluation is basic to all the rest of the material in the book and therefore requires careful attention.

1.3   The binary system of number representation

Most modern digital computers use the binary representation of numbers;

A typical decimal representation of a number:

$$437.69 \equiv 4 \times 10^2 + 3 \times 10^1 + 7 \times 10^0 + 6 \times 10^{-1} + 9 \times 10^{-2}$$

Binary representation:

$$1010.11 \equiv 1 \times 2^3 + 0 \times 2^2 + 1 \times 2^1 + 0.2 + 1 \times 2^{-1} + 1 \times 2^{-2}$$

---

that is, they represent numbers in the form of powers of 2 rather than of powers of 10.

examples

■ *The Conversion from One Number Representation to Another*

For integers, the method of conversion from decimal to binary is based on the observation that if the decimal representation is divided by 2 and there is a remainder of 1, then the last digit of the binary representation of the number is 1; otherwise it is 0. Repeated divisions of the quotients by 2 produces a sequence of remainders that is exactly the sequence of binary digits read from right to left.

For conversion from binary to decimal, we merely double the first digit on the left and add the next binary digit. We double again and add the next digit, and so on, which reverses the first process.

Example   *Convert decimal 418 to binary:*

$$
\begin{array}{l}
\qquad\qquad\qquad \textit{Remainder} \\
2\,\overline{)418} \\
2\,\overline{)209} + 0 \\
2\,\overline{)104} + 1 \\
2\,\overline{)52} + 0 \\
2\,\overline{)26} + 0 \\
2\,\overline{)13} + 0 \\
2\,\overline{)6} + 1 \\
2\,\overline{)3} + 0 \quad \textit{Read} \\
2\,\overline{)1} + 1 \quad \textit{digits} \\
\quad\ 0 + 1 \quad \textit{up}
\end{array}
$$

The answer is

$$110100010$$

Reverse process to go from binary to decimal:

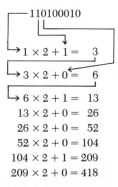

$$
\begin{array}{l}
110100010 \\
1 \times 2 + 1 = \quad 3 \\
3 \times 2 + 0 = \quad 6 \\
6 \times 2 + 1 = \quad 13 \\
13 \times 2 + 0 = \quad 26 \\
26 \times 2 + 0 = \quad 52 \\
52 \times 2 + 0 = 104 \\
104 \times 2 + 1 = 209 \\
209 \times 2 + 0 = 418
\end{array}
$$

The conversion of the decimal representation of numbers less than 1 to binary is based on repeated doublings and the use of the "carry" into the units position, in much the same way as the remainders were used for integers. The last possibility, conversion of the binary representation of numbers less than 1 to decimal, follows from the definition by multiplying the right-hand binary digit by $\frac{5}{10}$, adding the next binary digit (0 or 1), and again multiplying by $\frac{5}{10}$, and so on, until the last step, which ends with multiplying by $\frac{5}{10}$.

The processes we have described for the conversion from decimal to binary and back made use of decimal arithmetic; when the conversion is done inside the machine, different, though similar, processes which use binary arithmetic are necessary. These necessary changes are straightforward, once the basic idea of number representation is understood. Indeed, trouble in understanding binary representation arithmetic generally can be traced to a lack of understanding of how and why decimal arithmetic is done.

Convert decimal 0.314 *to binary*:

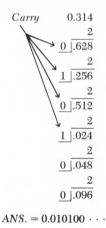

$$ANS. = 0.010100 \cdots$$

Routines for converting numbers from binary to decimal and back are generally included in the programming system used. But note that a terminating decimal like $0.1 = \frac{1}{10}$ cannot be written in a terminating binary representation, and hence adding — to itself 10 times will not produce the answer 1 inside a binary machine (just as in decimal representation arithmetic that uses a finite length of number representation, the sum $\frac{1}{3} + \frac{1}{3} + \frac{1}{3}$ does not equal 1).

$$Decimal \qquad Binary$$

$$0.1 = \frac{1}{10} = 0.0001100110011 \cdots$$

Because of the larger number of digits needed to write binary numbers, it is customary to use **octal** (base 8) numbers much of the time. The conversion to and from binary is a matter of grouping the binary digits into sets of three and using the corresponding octal digit for the triple of binary digits. There is currently a strong trend to use base 16 and hence a grouping of binary digits by fours.

CONVERSION TABLE

Binary	Octal
000	0
001	1
010	2
011	3
100	4
101	5
110	6
111	7

$$Binary\ 110100.010101 = (110)(100).(010)(101)$$

$$\downarrow \quad \downarrow \quad \downarrow \quad \downarrow$$

$$6 \quad 4\ .\ 2 \quad 5 \quad octal$$

The conversion routines supplied by the computing center usually have the peculiar feature that when a decimal number is read into the machine and is later printed out, a different number sometimes appears. Although the effect is all too often traceable to poor conversion routines, it is a basic fact that the conversion over and back cannot always reproduce the original number (even with a reasonable balance between the number of decimal places and the number of binary places).

Consider a conversion from three decimal to ten binary places and back. In particular, examine

$$
\begin{cases}
9.00 \longleftrightarrow 1001.\text{xxx xxx} \\
9.01 \longleftrightarrow 1001.\text{xxx xxx} \\
\vdots \\
9.99 \longleftrightarrow 1001.\underbrace{\text{xxx xxx}}
\end{cases}
$$

$\underbrace{}$

100 *numbers*   64 *numbers that*
in all   *are distinct*

Hence distinct decimal numbers must go into the same binary numbers and cannot be distinguished in the conversion back (even though $10^3 < 2^{10} = 1024$).

We shall not go further into the messy field of binary representation arithmetic and its pitfalls because much depends on the machine used, on the programs supplied by the computer manufacturer, and on the computer center, but it should be clear that the system of binary representation of numbers is going to affect our answers in many, often annoying, ways, especially when we least expect it.

PROBLEMS 1.3

In problems 1 to 9, convert the decimal representations to binary and check by reconverting:

1 1066
2 1492
3 1732
4 0.7071
5 0.5771
6 0.8660
7 3.14
8 2.718
9 1.732

Note that if a table of powers of 2 were available, then the conversion from binary to decimal could be reduced to a number of table-lookups followed by the necessary additions. Using this method in problems 10 to 12, convert the numbers from binary to decimal:

Binary number:

$$
\begin{aligned}
100100101 &= 2^8 + 2^5 + 2^2 + 1 \\
&= 256 + 32 + 4 + 1 \\
&= 293 \; decimal
\end{aligned}
$$

TABLE
<hr>

$2^1 = 2$
$2^2 = 4$
$2^3 = 8$
$2^4 = 16$
$2^5 = 32$
$2^6 = 64$
$2^7 = 128$
$2^8 = 256$
$2^9 = 512$
$2^{10} = 1,024$
$2^{11} = 2,048$
$2^{12} = 4,096$
$2^{13} = 8,192$
$2^{14} = 16,384$
$2^{15} = 32,768$
<hr>

10   1 010 010
11   11 001 100
12   1 010 101

Powers of 10:

	Binary
$10^0 \longrightarrow$	1
$10^1 \longrightarrow$	1010
$10^2 \longrightarrow$	1100100
$10^3 \longrightarrow$	1111101000

In problems 13 to 15, use the table to convert the numbers from decimal to binary by subtracting the appropriate powers of 2:

13 1620
14 1588
15 1968
16 Discuss the use of tables for the conversion of numbers less than 1 in size.
17 How would the table-lookup method of conversion work inside a binary machine?
18 Can a terminating binary representation not terminate in decimal form? Compare the respective numbers of digits in the two forms.
19 Describe the terminating decimal representation numbers which are not terminating binary numbers.
20 Show that the argument that a three-decimal to ten-binary conversion cannot always be unique can be adapted to the widely used eight-decimal to twenty-seven binary conversion and back.
21 Show that "Halloween" is the same as "Christmas," that is, OCT 31 = DEC 25.

1.4   The three number systems

Besides the various bases for representing numbers—decimal, binary, and octal—there are also three distinct number systems that are used in computing machines.

First, there are the integers, or counting numbers, that are used to index and count. Usually, they have the range from 0 to the largest number that can be contained in the machine's index registers.

Typically, the range of counting numbers is 0, 1, 2, 3, . . ., 32,767

Second, there are the fixed point numbers whose length is usually (sometimes a simple multiple of) one word of the machine. Although nominally we can imagine the decimal point (or binary point if you wish) to be at either end of the number, the user can imagine that it is any place he pleases and make compensation for the machine's behavior by using the appropriate shift operations.

Typical fixed point numbers:

$$376.476\ 392\ 7$$
$$-593,762.4716$$
$$-0.001\ 117\ 179$$

Here the burden is clearly put on the user to determine the range of the numbers computed by the program he writes. We usually have little difficulty in estimating the maximum size of either a sum or a product of numbers; the difficulty arises when we want to divide. In division we must estimate the *minimum* possible value of the denominator in order to estimate the maximum of the quotient, and the estimation of a minimum is very difficult because of the possibility of almost complete cancellation in adding two numbers of opposite sign and approximately equal size.

The fixed point number system is the one that the programmer has implicitly used during much of his own calculation, and it is the one with which he is most familiar. Perhaps the only feature that is different in hand and machine calculations is that the machine always carries the same number of digits, whereas in hand calculation the user often changes the number of figures he carries to fit the current needs of the problem.

Third, there is the floating point number system which is the one used in almost all practical scientific and engineering computations.

Typical floating point numbers:

$$\pi = 0.314 \times 10^1$$
$$100\pi = 0.314 \times 10^3$$
$$\frac{\pi}{1000} = 0.314 \times 10^{-2}$$

$$\uparrow \qquad \uparrow$$
$$mantissa \quad exponent$$

This number system differs in significant ways from the fixed point number system, and we must be aware of these differences at many stages of a long computation. Typically, the computer word length includes **both** the mantissa and the exponent; thus the number of digits in the "mantissa" of a floating point number is less than in that of a fixed point.

1.5 Floating point numbers

The floating point numbers are not uniformly spaced (as are the numbers in the mathematician's real number system), but are "bunched up" around the origin. For convenience in exposition and for following the examples, if we assume that we have three decimal

---

The number system we shall use most of the time for illustration has

Three-decimal-place mantissas
Exponent range of −9 to 9

$$\sqrt{2} = 0.141 \times 10^1$$
$$\frac{-1}{\sqrt{3}} = -0.577 \times 10^0$$
$$10 = 0.100 \times 10^2$$

---

places in the mantissa and one in the exponent (instead of 8 to 10 decimal places in the mantissa and an exponent range of around ±38 as is usual in computing machines), we can easily see what happens and why it happens. (The reason for placing the decimal point **before** the first digit arose from fixed point arithmetic and is not relevant to floating point arithmetic, but we shall follow the custom and place it there.)

For example, there are as many numbers between 0.100×10^{-3} and 0.999×10^{-3} as there are between 0.100×10^4 and 0.999×10^4 or any other decade of numbers.

The number 0 occupies a unique position in the number system, and designers of the computers have had a great deal of trouble with it.

The number that is "next to zero" is 0.100×10^{-9}.
The next number is 0.101×10^{-9}.
*Hence, the finest spacing is 10^{-12} and **does not occur adjacent to zero**.*
The largest number is 0.999×10^9.
The adjacent number is 0.998×10^9.
Hence, the spacing is 10^6 at the large end of the number system.

There is, for example, the question of how to cope with both $+0$ and -0. Again, in floating point if a number is subtracted from itself, what is to be the answer? The numerical analyst would like to have (most of the time) 0.000×10^{-9}. It takes both time and equipment to get the exponent right, but failure to do so may cause trouble at a later stage.

We can say that floating point numbers are something like logarithms since they are bunched up around zero, but since both positive and negative numbers occur in the floating point

---

*The characteristic of a log is the same as the characteristic (exponent) of a floating point number, **but mantissas are not the same**.*

Floating point system has "geometric progression spacing" due to the exponent but "arithmetic progression spacing" due to the mantissa.

---

number system, they are not the same. Thus, we can say that the floating point number system is distinctly different from the usual number systems in mathematics, and we must again warn the student that these differences can produce surprising results at times.

The finite range of the exponent also is a source of trouble, namely, what are called "overflow" and "underflow," which refer respectively to the numbers exceeding the largest- and the smallest-sized (non-zero) numbers that can be represented within the system.

It should be evident that we can replace an underflow by a zero and often not go far wrong. It is less safe to replace a positive overflow by the largest number that the system has, namely, 0.999×10^9 (or even 0.900×10^9 to prevent **some** subsequent overflows due to future additions). For this reason, we shall tend to prefer forms of organizing the arithmetic that are more vulnerable to underflow than to those that are vulnerable to overflow.

We may wonder how, in actual practice, with a range of 10^{38} to 10^{-38} or more, we can have trouble with overflow and underflow. The answer is quite simple. Very frequently, there is an adequate theory for small values of some parameter, and there is another theory for large values of the same parameter, and the computer is used to explore from the region where some terms are very small (or dominating)[1] to the region where some other terms are very small (or dominating).

When is

$$e^x > 0.999 \times 10^9$$

Say 10^9*. Taking logs,*

$$x > 9 \ln 10$$
$$> 20.7$$

Hence, if

$$x > 21$$

then there will be an overflow when e^x *is calculated.*

For underflow,

$$x < -21$$

Thus, it is quite common to have the computer do problems having a very large range of parameter values; but even for a modest range of x, a function like e^x has a very large range. For example, in computing

$$1 - \frac{1}{e^x + 1}$$

for modest-sized positive x, we may get overflow when we compute e^x. If we rewrite the expression as

$$\frac{1}{1 + e^{-x}}$$

then we shall get underflow for the exponential, and its replacement by 0.000×10^{-9} will not lead us far astray.

---

[1]In this notation *always* read the word before the parentheses or else the word in the parentheses.

The function

$$f(x) = \frac{x^2 e^x}{e^x - 1} \qquad 0 \le x \le 100$$

will produce overflow in e^x at $x > 21$. But rearranging

$$f(x) = \frac{x^2 e^{-x}}{1 - e^{-x}}$$

will produce underflow, and the result is reasonable, if not always exactly correct.

PROBLEMS 1.5

1 How many numbers are there in the number system we are using (assuming a single zero)?
2 Explain the remark, "The number 0 is relatively isolated from its neighbors."
3 Write a short essay comparing a floating point number system with logarithms.
4 In the floating point binary notation we could, in principle, not store the first digit since, except for the number 0, this digit is always 1. Discuss the details of this proposal.

1.6 An example: Checking a fixed point sine table

As an example of the difference in spirit between fixed point and floating point arithmetic, consider Euler's formula for checking the correctness of a (fixed point) sine table

$$\sin A = \sin (36° + A) - \sin (36° - A) - \sin (72° + A) + \sin (72° - A)$$

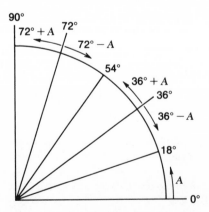

As A goes from 0 to 18°, each of the other terms covers a range of 18° (at the same spacing of values), and in total the whole range of 0 to 90° is uniformly covered. But in a floating point sine routine we expect that

$$A = 10^{-6}$$

will give (for A in radians)

$$\sin A = 10^{-6}$$

accurately. Such a value would not be well checked by Euler's formula; indeed, the spirit of the formula is wrong for floating point arithmetic since the additions would conceal relatively large errors that might occur in small angles.

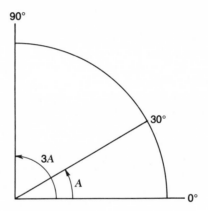

Instead of using Euler's formula, one is inclined to use the identity

$$\sin 3A = \sin A(3 - 4 \sin^2 A)$$

Here, as A goes from 0 to 30°, $3A$ goes from 0 to 90°. Since

$$4 \sin^2 A \leq 1 \qquad (0 \leq A \leq 30°)$$

then in computing

$$3 - 4 \sin^2 A$$

there is no great loss of accuracy owing to cancellation in the leading digits and the consequent shifting to "renormalize" the floating point number. The final multiplication also does not introduce large errors. Thus, we can easily check very small angles as well as moderate-size angles.

PROBLEMS 1.6

1 Derive the formula for tan $3x$ in terms of tan x, and discuss its possible use in checking a floating point tan x routine.

2 Discuss the formula

$$\cos 3x = 4 \cos^3 x - 3 \cos x$$

for checking a cos x table.

3 Generalize Euler's formula to n terms.

1.7 Roundoff error

When two 3-digit numbers are multiplied together, their product has either five or six places. If we are to keep only three places, then we must drop two or three places in the product.

Multiplication

$$0.236 \times 10^1$$
$$\times 0.127 \times 10^1$$
$$\overline{1652}$$
$$472$$
$$236$$
$$\overline{0.0299\,|72 \times 10^2}$$
$$0.\,|5\;\;\textit{roundoff}$$

ANS. $\underline{0.300}\;\;\textit{(drop)} \times 10^1$

$$|\textit{Roundoff error}| = 0.28 \times 10^{-2}$$

Assuming that the product is positive, good practice is to round, that is, to add a 5 (before dropping), to the first place that is to be dropped. The result is that if this place is 5 or more, then there is an increase in the last digit kept, but if it is less than 5, then there is no change in the last digit kept. This small change is called "roundoff error" and is an inevitable result of using a finite length of number representation. Unfortunately, many machines drop without rounding, which is called "chopping"; this can cause serious trouble.

Chopping

As above, the product is

$$0.0299\,|72 \times 10^2$$
$$\llcorner\;\textit{chop}$$

ANS. 0.299×10^1

A similar effect occurs in division. There can also be a roundoff in addition, even when the exponents of the two numbers are the same, provided there is a "carry" on the extreme left.

Addition

$$
\begin{array}{r}
0.742 \times 10^2 \\
+0.927 \times 10^2 \\
\hline
1.669 \times 10^2 \\
5 \quad \textit{roundoff} \\
\hline
0.167 \quad \times 10^3
\end{array}
$$

To a person familiar only with hand calculation, which rarely exceeds five decimal places, roundoff in an eight or more decimal machine seems like a trivial nuisance, but when we recall that we may commit these small errors on each of the many, many operations the machine is going to do, then it is possible to see why roundoff error occupies such a large part of numerical analysis.

Roundoff causes trouble mainly when two numbers of about the same size are subtracted.

$$
\begin{array}{r}
0.314 \times 10^1 \\
-0.313 \times 10^1 \\
\hline
0.001 \times 10^1 = 0.100 \times 10^{-1}
\end{array}
$$

As a result of the cancellation of the leading digits, the number is shifted (normalized) to the left until the first digit is not zero. This shifting can bring the roundoff errors that were in the extreme right part of the number well into the middle, if not to the extreme left. Later steps (and conversion to decimal) may well conceal the trailing zeros that were brought in by the shifting, so that we shall think that we have an accurate number when we do not. It is not possible in practice to foresee all the consequences of an extensive calculation, and we must therefore be constantly on our guard against assuming that the neatly printed sheets from the computer always give accurate, dependable numbers.

A second, more insidious trouble with roundoff, and **especially** with chopping, is the presence of internal correlations between numbers in the computation so that, step after step, the small error is always in the same direction and one is therefore not under the protective umbrella of the statistical "average behavior."

1.8 Two views of the numbers in a computer

One view of the numbers arising in a practical computation is that each true mathematical number is "smeared" a bit around the proper place, and that we cannot be sure just where in the smear the machine number is.

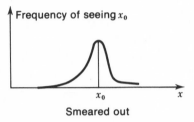

Smeared out

The words "a ton of sand" usually do not mean exactly 1 ton down to the last grain of sand, but rather an amount reasonably near 1 ton, where "reasonably near" depends upon circumstances. When the contractor delivers a ton of sand, he does not even attempt to get it correct within 1 pound; rather he expects to miss by tens of pounds, up to perhaps even 100 pounds. After all, the weight will change a moderate amount depending on the moisture in the sand. As remarked, this "smear" or "fuzzy" view is very useful at times in physical problems.

In mathematics "zero" occurs in two relationships

$$a - a = 0$$

$$a \cdot 0 = 0$$

In computing, these are slightly different. The "multiplicative zero"

$$0.000 \times 10^{-9}$$

does indeed behave properly,

$$0.314 \times 10^1 \times 0.000 \times 10^{-9} = 0.000 \times 10^{-9}$$

---

but the "additive zero" that can occur from

Is $0.100 \times 10^5 - 0.100 \times 10^5 = 0.000 \times 10^{-9}$?

Or perhaps 0.500×10^2?

---

the roundoff in such operations as

$$(0.100 \times 10^3)(0.100 \times 10^3) - (0.990 \times 10^2)(0.101 \times 10^3)$$
$$= 0.100 \times 10^5 - 0.100 \times 10^5 = 0$$

can lead us astray, because we think that we have the exact zero, 0.000×10^{-9}.

A second view boldly states that the only numbers are those that can be represented in the machine—there are no other numbers. In a sense this view states that the mathematician's real number system is completely fictitious,

*Above is a "picture" of the **only** numbers there are—there are no others!*

and that the number system the machine uses (and the user sees) is the "real number system." Although this view may appear to be a bit extreme, it has some merit; it is especially useful in debugging programs when it is sometimes necessary to account for exactly how each number was obtained, down to the last digit. It is also sometimes useful when merely thinking about a computation and wondering what will happen (or has happened).

example

■ *Explain why some values of tan x cannot be obtained when x is between 0 and π/4.*

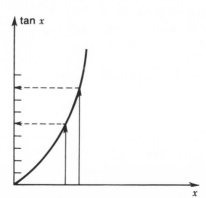

The slope of $y = \tan x$ is $\sec^2 x$, and this is larger than 1 for all values of $x \neq 0$. Thus, there will be intervals where adjacent values of x do not lead to adjacent values of $\tan x$. Indeed, the larger x is, the more there will be gaps in the output because of the "granularity" of the input numbers x.

PROBLEMS 1.8

1 Discuss the relative merits of arctan x versus arcsin x library routines.
2 Show that sin x takes on consecutive values as x goes through the interval $0 \leq x \leq \pi/2$, but does not for the interval $0 \leq x \leq \pi$.
3 Discuss in terms of Sec. 1.8 the problem of evaluating sin x for large x.
4 Discuss the trouble in computing cos x for x near $\pi/2$.

1.9 Relative error

The floating point notation tends to put emphasis on "relative error"

$$\frac{\text{True} - \text{calculated}}{\text{True}}$$

rather than on the "absolute error"

$$\text{True} - \text{calculated}$$

This effect tends to agree with practice, where it is usually the error in the number **relative** to the size of the number that is important, rather than the size of the error itself. Relative error has meaning independent of the units of measurement, whereas the absolute error depends on the units used for the measurement.

Relative error in calculating sin π is

$$\frac{0 - \text{calc}\,[\sin{(0.314 \times 10^1)}]}{0}$$

and it is unlikely that

$$\sin{(0.314 \times 10^1)} = 0.000 \times 10^{-9}$$

However, the relative error is not always suitable. For example, when the number being computed is theoretically zero, such as sin π, then any error in the answer would produce an infinite relative error. Still using the sin x as an example, consider trying to compute sin

($\pi \times 10^6$), which mathematically should be zero. In our quantized number system in the computer, the distance between the successive numbers 0.314×10^7 and 0.315×10^7 is so large that the mathematical function will go through many cycles in passing from one number to another. What would be a reasonable answer in this situation? Giving the midrange value, zero, of the sine would surely be misleading.

Perhaps for large angles we should use

$$Maximum \ \{|fcn|,|argument|\}$$

as the denominator.

The failure to recognize the importance of thinking in terms of relative error leads people into simple mistakes. Unfortunately, the usual mathematics course stresses the absolute error so much that it is almost automatic for the beginner to reason in terms of the absolute rather than the relative error.

example

■ *Calculate e^{-x} for $0 \leq x \leq 3$.*

One method is to use the first k terms of the Taylor expansion of e^{-x} about $x = 0$; another method is to use the expansion of e^x and then divide 1 by it to get e^{-x}. The careless student will reason somewhat as follows: How many terms do I need? Since he knows that the terms in the expansion of e^{-x} will become monotone decreasing and alternating in sign after a while, he sets the $(k + 1)$th term less than the desired error, say 10^{-3}, and solves approximately for k. However, note that we shall not have three significant figures in all cases because $e^{-3} \simeq 0.050$ is rather small.

$$e^{-x} = 1 - x + \frac{x^2}{2} - \cdots + \frac{(-1)^k x^k}{k!} + \cdots$$

The condition

$$\left| \frac{x^{k+1}}{(k+1)!} \right| \leq 10^{-3} \qquad 0 \leq x \leq 3$$

requires

$$3^{k+1} \leq (k+1)! \times 10^{-3}$$

or

$$k \approx 12$$

Hence, 12 terms.

The relative error for e^x requires about nine terms.

The expert will recognize that using the expansion of e^x, even though the truncation error will be larger than for e^{-x}, will require fewer terms **relative** to the size of the answer (namely, e^x) and that the final division $1/e^x = e^{-x}$ will not introduce serious further errors.

Here we have a simple example of how the methods learned in the usual mathematics courses tend to lead the person in computing astray. It is not that mathematics is wrong in emphasizing the absolute error, but rather that computing tends to emphasize **relative** error and the two approaches require different thinking.

While it is not true, as we have noted, that the relative error is the appropriate one to use in **all** cases, we will tend to emphasize it repeatedly. One of the results of this emphasis is that many of our methods will differ from those that have been highly recommended in the past; the author believes that error analysis has too often centered around the absolute error, and that in practice the absolute-error estimates are usually far more difficult to understand and interpret back to the original physical situation than are the relative-error estimates.

PROBLEMS 1.9

1 Discuss in some detail the suggestion in the italic insert that the error should be relative to

$$\max \{|x|, |f(x)|\}$$

2 Discuss the advantage of having both $\log x$ and $\log (1 + x)$ in the library.

1.10 Zeros of a function

In mathematics the idea of a zero of a function is a simple one, but in computing one can fairly ask: What is wanted of the zeros of a polynomial $P(x)$?

[1] That those values x_i which make $|P(x_i)|$ small should be as accurate as required?

[2] That at the values x_i, the polynomial $P(x_i)$ should be as small as required?

[3] That the polynomial may be reconstructed as accurately as required from the zeros.

Given

$$x^2 + 80x + 1 = 0$$

The standard formula gives

$$x_i = \frac{-b \pm \sqrt{b^2 - 4ac}}{2a}$$

$$= \frac{-0.800 \times 10^2 \pm \sqrt{0.640 \times 10^4 - 0.400 \times 10^1}}{0.200 \times 10^1}$$

$$= \begin{cases} -0.800 \times 10^2 & \textit{if} + \textit{used} \\ 0.000 \times 10^{-9} & \textit{if} - \textit{used} \end{cases}$$

But note: If we use

$$\begin{cases} x_1 = - \, sign \, (b) & \left\{ \dfrac{|b| + \sqrt{b^2 - 4ac}}{2a} \right\} \\ x_2 = \dfrac{c}{ax_1} \end{cases}$$

then

$$x_1 = -0.800 \times 10^2$$
$$x_2 = -0.125 \times 10^{-1}$$

Also note:

$$(x - x_1)(x - x_2) = x^2 + 80x + 1$$

as it should!

What we want depends on the particular problem and cannot be answered definitively in advance. Careful consideration shows that often the answer is [2] but more often it is [3], whereas at first glance the beginner thinks that it is [1].

PROBLEMS 1.10

Using our three-digit arithmetic:

1 Compute the zeros of

$$x^2 + 100x - 4 = 0$$

2 Compute the zeros of

$$x^2 - 60x + 2 = 0$$

3 Derive the formula for the zeros of

$$Ax^2 + 2Bx + C = 0$$

and find the savings in the arithmetic to be done.

1.11 Function evaluation

Having examined some of the details of the machine and of the numbers that we are going to use in computing, let us look at what we are going to compute.

Probably the most basic computing is that of function evaluation. Furthermore, function evaluation is fundamental to much of the rest of computing. Therefore, for the rest of this chapter we shall, in one way or another, concern ourselves with this apparently simple process.

Unfortunately, even in simple function evaluation, there may be a severe loss of accuracy owing to the cancellation of almost equally sized numbers. Thus, if

$$f(x) = \sqrt{x+1} - \sqrt{x}$$

and x is small, there is no trouble; but when x is large, there is severe cancellation.

In evaluating

$$\sqrt{x+1} - \sqrt{x}$$

for large x, rearrange as follows:

$$(\sqrt{x+1} - \sqrt{x})\,\frac{\sqrt{x+1} + \sqrt{x}}{\sqrt{x+1} + \sqrt{x}} = \frac{1}{\sqrt{x+1} + \sqrt{x}}$$

Although there appear to be a large number of tricks to handle this kind of a situation, they are not new to the student—the tricks are exactly the same as those used in the calculus to evaluate

$$\lim_{\Delta x \to 0} \left\{ \frac{\Delta y}{\Delta x} \right\} = \lim_{\Delta x \to 0} \left\{ \frac{f(x + \Delta x) - f(x)}{\Delta x} \right\}$$

In evaluating

$$\frac{\sin (x + \Delta x) - \sin x}{\Delta x} = \cos \left(x + \frac{\Delta x}{2}\right) \left[\frac{\sin (\Delta x/2)}{\Delta x/2}\right] \quad for\ small\ \Delta x$$

Also, for small Δx,

$$1 - \cos \Delta x = \frac{\sin^2 \Delta x}{1 + \cos \Delta x}$$

$$= 2 \sin^2 \frac{\Delta x}{2}$$

All that is required is that the person computing should use his imagination and foresee what might happen **before** he writes the program for a machine. As a simple rule, try to avoid subtractions (even if they appear as a sum but with the sign of one of the terms negative and the other positive).

$$(x + \varepsilon)^{2/3} - x^{2/3} = [(x + \varepsilon)^{2/3} - x^{2/3}] \frac{(x + \varepsilon)^{4/3} + (x + \varepsilon)^{2/3} x^{1/3} + x^{4/3}}{(x + \varepsilon)^{4/3} + (x + \varepsilon)^{2/3} x^{2/3} + x^{4/3}}$$

$$= \frac{[(x + \varepsilon)^{2/3}]^3 - (x^{2/3})^3}{(x + \varepsilon)^{4/3} + (x + \varepsilon)^{2/3} x^{2/3} + x^{4/3}}$$

$$= \frac{2x\varepsilon + \varepsilon^2}{(x + \varepsilon)^{4/3} + (x + \varepsilon)^{2/3} x^{2/3} + x^{4/3}}$$

The examples in the italic inserts of the page are only a few of the methods that the student has used in another context to rearrange expressions so that the result can be evaluated on a computer reasonably accurately.

PROBLEMS 1.11

How would you evaluate (ε small)?

1. $\dfrac{1}{x + 1} - \dfrac{1}{x}$

2. $\cos (x + \varepsilon) - \cos x$

3. $\sqrt[3]{x + 1} - \sqrt[3]{x}$

4. $1/\sqrt{x} - 1/\sqrt{x + 1}$

5. $\tan (x + \varepsilon) - \tan x$

6. $\dfrac{1}{x + 1} - \dfrac{2}{x} + \dfrac{1}{x - 1}$

1.12 Other methods for rearranging an expression

Simple rearrangements will not always produce a satisfactory expression for computer evaluation, and it is necessary to use other devices that occur in the calculus course.

For small positive x:

$$1 - e^{-x} \equiv x - \frac{x^2}{2} + \frac{x^3}{3!} - \cdots$$

$$\ln(1 - x) = -\left(x + \frac{x^2}{2} + \frac{x^3}{3} + \cdots\right)$$

One of the most useful devices for obtaining a suitable expression for evaluation is to use a finite number of terms of a power series expansion. A number of examples are shown in the inserts.

Example

Evaluate for small x,

$$\frac{\tan x - \sin x}{x^3} = \frac{\left(x + \frac{x^3}{3} + \frac{2x^5}{15} + \cdots\right) - \left(x - \frac{x^3}{6} + \frac{x^5}{120} + \cdots\right)}{x^3}$$

$$= \frac{\left(\frac{1}{3} + \frac{1}{6}\right)x^3 + \left(\frac{2}{15} - \frac{1}{120}\right)x^5 + \cdots}{x^3}$$

$$= \frac{1}{2} + \frac{x^2}{8} + \cdots$$

Another technical device of less practical use but of great theoretical value is the mean-value theorem

$$f(b) - f(a) = (b - a) f'(\theta) \qquad (a < \theta < b)$$

Whereas the value of θ is not known and in principle can be anywhere inside the interval (a, b), it is reasonable to suspect that the choice of the midvalue is as good as any other value **if nothing else is known about the function.**

For x small with respect to a ,

$$\ln (a + x) - \ln a = \ln \left(1 + \frac{x}{a}\right)$$

Also by using the mean-value theorem,

$$\ln (a + x) - \ln a = \frac{x}{a + \theta}$$

$$\simeq \frac{x}{a + x/2}$$

Thus in **estimating** an expression, often the midvalue of the interval is used; in **bounding** an expression, the maximum of $f'(\theta)$ in the interval is used.

PROBLEMS 1.12

How would you evaluate Problems 1 to 3 for small x?

1 $\dfrac{1 - \cos x}{x^2}$

2 $\dfrac{xe^{(a+b)x} - 1}{(e^{ax} - 1)(e^{bx} - 1)}$ $\left(\text{One answer, } \dfrac{a + b}{ab}\right)$

3 $\ln \dfrac{1 - x}{1 + x}$

4 Discuss the "chain method" of evaluating a polynomial
$$P(x) = a_n x^n + a_{n-1} x^{n-1} + \cdots + a_0$$
$$= [\cdots (a_n x + a_{n-1})x + \cdots]x + a_0$$

1.13 An example: Computing the sum of a large number of numbers

There are a number of different tricks that can be used to decrease the effect of roundoff; the main problem is to discover when roundoff is giving trouble and where in the total calculation it is causing the most trouble. As an example of the kind of tricks that are comparatively easy to invent once the cause is understood, consider the problem of adding (using single-precision floating point arithmetic) a large number of floating point numbers.

If we simply add them one at a time, there will generally be a lot of roundoff accumulated. One way of avoiding this is to sort the numbers into two columns, one column being the positive numbers and the other the negative, with each ordered in size from large to small. We now begin to add, taking the smallest in one column, say the positive column, and subtract (add) the negative ones (starting at the smallest) until the sign of the sum changes. Each time the sign changes, we go to the other column to find the next number, which keeps the accumulated sum as close to zero as possible. When finally one column is exhausted, the remaining numbers in the other column are included in the total.

Obviously, this method will be expensive in sorting effort, and it is reasonable to ask if simply doing the whole in double-precision arithmetic (which is generally included in the basic operations of modern machines) would not generally be preferable. As has often been said, "The answer to roundoff troubles is simply to use double precision."

PROBLEMS 1.13

  1 Discuss why the two-column method might be good.
  2 Discuss the minor improvement obtained by examining the two columns to find exactly equal terms and cancelling them **before** starting the summation.

1.14   Using the machine to preserve accuracy

The problem of computing expressions like

$$e^{ax} - 1$$

for small $|ax|$ arises often and may cause a serious loss of accuracy if e^{ax} is computed first. The usual approach is to try to analyze those expressions in which this phenomenon will occur and to compute them by, say, a truncated Maclaurin series

$$ax + \frac{(ax)^2}{2!} + \frac{(ax)^3}{3!} + \frac{(ax)^4}{4!} + \frac{(ax)^5}{5!} + \cdots$$

rather than by direct formula.

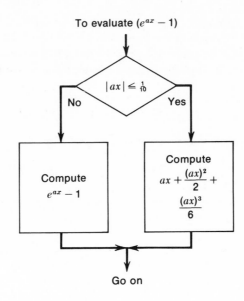

To evaluate $(e^{ax} - 1)$

$|ax| \leq \frac{1}{10}$

No   Yes

Compute
$e^{ax} - 1$

Compute
$ax + \dfrac{(ax)^2}{2} + \dfrac{(ax)^3}{6}$

Go on

A far better approach is to let the machine do the analysis by simply testing for small $|ax|$ and branching accordingly.

The same idea applies in many other situations where there may or may not be a serious loss of accuracy, depending on the particular values used; let the machine test and branch to the appropriate process.

PROBLEM 1.14

1   Apply machine decision method to $\ln(1 + x)$.

1.15   Types of computing

In the practice of computing, many different situations arise. The simplest are in computing the answer for isolated, one-shot problems in which it is rather well known how the computation will go.

In the typical one-shot problems, machine time is not important.

A second kind of situation occurs when computing many cases (some of which have not yet been thought of but will develop as the earlier ones are done). Unfortunately the new computations are often far outside the range originally considered reasonable, and it is necessary to apply some imagination **before** writing down the details of the computing (if the program is not to be rewritten several times).

Many computing problems require reasonable attention **before** *starting to program.*

A still more difficult problem is writing a library program for general use. Here, extreme care is needed to cover as many situations as is reasonable without writing such an inefficient or complex program that it will not be used very often.

Library routines require great care.

A careful treatment of "error returns" and exceptional cases is required. We shall give closer attention to this difficult topic in Chap. 15.

Sometimes the problem is so large and will use so much machine time that it is necessary to make a careful examination of the particular class of problems to be solved, and to take advantage of the "physical intuition" of the man who proposed the problem, before a practical routine can be designed.

Large, time-consuming problems often need methods appropriate to the special class of problem being solved.

This class of problems requires an intimate acquaintance with the field of application and lies outside the scope of this book.

A final type of problem is the "real-time" problem as in a program for a space vehicle, a guided missile, or a process control, when it is essential to run the problem within a fixed length of time.

Real time problems often require great care in preparation.

As in the preceding problem, we can afford to use much time and effort in the preliminary examination and design of the formulas to be computed, and again this lies outside the scope of the book.

Another way of saying the same thing is that in computing there is a need for [1] "handy-dandy," readily available though possibly inefficient methods; [2] carefully thought-out methods which do not take advantage of a particular situation; and [3] special situations in which the details of the application can be very important. Only types [1] and [2] are examined in this book.

The attitude adopted in this book is that while we expect to get numbers out of the machine, we also expect to take action based on them, and, therefore we need to understand thoroughly what the numbers may, or may not, mean. To cite the author's favorite motto,

"The purpose of computing is insight, not numbers,"
although some people claim,

"The purpose of computing numbers is not yet in sight."

There is an innate risk in computing because "to compute is to sample, and one then enters the domain of statistics with all its uncertainties."

1.16 Theories of roundoff

There are a number of different approaches to roundoff. We have throughout the book chosen to emphasize avoidance, rather than theories, of roundoff. Most books show how roundoff is propagated through the four simple arithmetic operations and then go on to the important compound operation of the vector dot product of two vectors **x** and **y**

$$x_1 y_1 + x_2 y_2 + \cdots + x_n y_n$$

The main tool for this kind of estimation is the differential calculus (in the form of differentials, which is frowned upon in some pure mathematics books).

$$\frac{\partial f(x_1, x_2, \ldots, x_n)}{\partial t} \Delta t = \frac{\partial f}{\partial x_1} \Delta x_1 + \frac{\partial f}{\partial x_2} \Delta x_2 + \cdots + \frac{\partial f}{\partial x_n} \Delta x_n$$

We often need to estimate the roundoff that will occur **before** we begin a calculation since we want to know before we start if we shall have to use double precision to get the accuracy we want.

Range (interval) arithmetic makes valuable use of the machine **during** a computation to get bounds on the roundoff error. The idea behind this method is that every number is replaced by a pair of numbers, called the **range** (interval). Each arithmetic operation becomes the computation of the new range of the result. This is nontrivial and occupies a lot of machine time. On short problems, range arithmetic is very useful.

There is also the problem of estimating the roundoff **after** a computation is done. This type of estimation is valuable because it is in many ways independent of the particular result being computed. In Chap. 6 we shall examine the estimation of the roundoff in a table of numbers.

We have indicated that there are both error bounds, as in range arithmetic, and estimates of the probable roundoff error. In small problems the bounds are very suitable, but in large problems such as trajectories to the moon the bounds can be so pessimistic that from a practical point of view we must be satisfied with statistical estimates of the roundoff effects (which is unpopular with mathematicians and so has been too little studied).

Another type of error analysis is the so-called backward analysis which gives answers like, "You have the exact answer to a problem that is not more than so far from your original problem." Sometimes this type of answer can be very useful, but sometimes it is in a form that the user cannot take advantage of, and, if so, it should be avoided.

We have emphasized, and shall continue to emphasize, methods that tend to avoid roundoff because in a first course they seem to be most useful. But this should not be interpreted to mean that bounds and estimates of roundoff are not important.

REAL ZEROS OF A FUNCTION

2.1 Introduction

The problem of finding the real zeros of a given continuous function arises frequently in science and engineering. In the present chapter we shall examine this problem, and in the next chapter we shall examine that of finding the complex zeros of an analytic function and shall leave the important discussion of polynomials with its special methods to Chap. 4.

The mathematician considers a zero of a function

$$y = f(x)$$

as any number x such that when it is substituted in the function $f(x)$, the result is exactly zero.

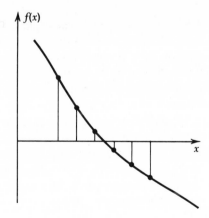

However, in practical computing, as we observed in Chap. 1, there is only a discrete set of numbers to try as possible zeros, and none of these may happen to produce exactly zero for $f(x)$. We shall therefore seek a pair of numbers, x_1, x_2, which are "close to each other" and such that $f(x_1)$ and $f(x_2)$ have opposite signs (or in rare cases one of them may be zero).

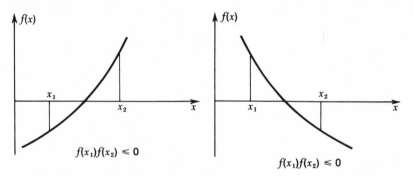

$$f(x_1)f(x_2) \leq 0$$

In practice, because of the random effects of the roundoffs which occur in the function evaluation of $f(x)$, we usually have not a single change of sign, but a short sequence of consecutive numbers which gives a sequence of changes in sign for $f(x)$.

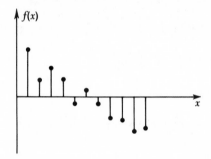

Although in principle this could be confused with a sequence of distinct zeros, in practice there is little trouble with this effect **provided** we can form a reasonable estimate of the size of the roundoff error made during the function evaluation.† In order to reduce this roundoff error we need, obviously, to practice the techniques discussed in Chap. 1.

Roundoff troubles are bound to occur in the problem of finding zeros because it is in the nature of the problem that there is almost an exact cancellation between the positive terms and the negative terms that occur in the function evaluation process.

Often the function to be examined has an infinite number of real zeros, for example,

$$f(x) = \tan x - x \tag{2.1}$$

---

†See Sec. 6.15.

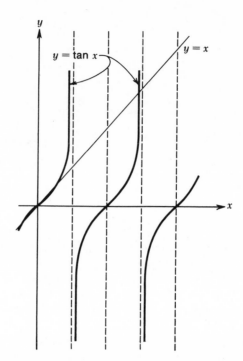

and we must be satisfied with finding only those zeros which lie in some given interval. We must expect that although we can determine the zeros near the origin quite accurately, those far away will be determined poorly in absolute accuracy, but we can hope that all the zeros will have about the same relative accuracy (excluding the special $x = 0$).

PROBLEMS 2.1

1 Using a graph, find the zero near $x = 3\pi/2$ to two decimal places for

$$\tan x = x$$

2 Find the first two figures of the first two real, positive zeros of $x \sin x = 1$.

3 Estimate to two places the real zero of

$$xe^x = 1$$

4 Show that there are no real zeros of

$$e^x = 1 + \ln x$$

by examining their graphs.

5 Find the smallest positive solution of

$$\tan x = x^2 + 1$$

to two decimal places.

2.2 The bisection method

The **bisection method** is the simplest method for finding the real zeros of a continuous function $f(x)$ in an interval $a \leq x \leq b$ and is probably the safest if you are going to depend completely on a computer routine. The heart of the bisection method is the **assumption** that an interval $x_1 \leq x \leq x_2$ has been found such that

$$f(x_1)f(x_2) < 0$$

and the method undertakes to decrease the size of the interval. This decrease is accomplished by evaluating the function $f(x)$ at the midpoint of the interval, that is, computing

$$f\left(\frac{x_1 + x_2}{2}\right)$$

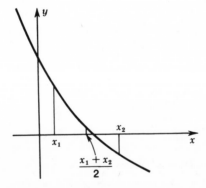

If this value is not zero, then there is a sign change in one of the two halves of the original interval and we pick the new, decreased interval with the sign change by

$$f(x_1)f\left(\frac{x_1+x_2}{2}\right) = \begin{cases} 0 & \text{a zero at } \dfrac{x_1+x_2}{2} \\[2mm] <0 & \text{new interval } \left(x_1, \dfrac{x_1+x_2}{2}\right) \\[2mm] >0 & \text{new interval } \left(\dfrac{x_1+x_2}{2}, x_2\right) \end{cases}$$

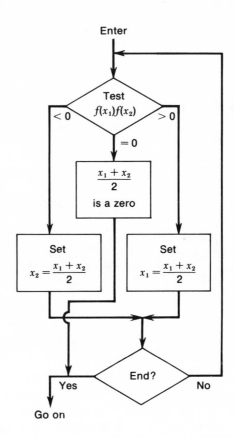

Thus, $\frac{1}{2}(x_1+x_2)$ replaces either x_1 or x_2 (most of the time), and we can repeat the process again and again. Each repetition halves the length of the interval, and 10 repetitions, for example, reduce the length of the original interval by a factor of

$$2^{10} = 1024 > 1000 = 10^3$$

Thus 10 or, at most, 20 repetitions are all that are likely to be required.

Find

$$x = \sqrt{2}$$

The number is a zero of

$$y = x^2 - 2$$

Start

$$y(1) = -1$$
$$y(2) = +2$$
$$y(1)y(2) < 0$$

Range

$$1 \le x \le 2$$

Bisect

$$x = 1.5$$
$$y(1.5) = 0.25$$
$$y(1)y(1.5) < 0$$

Choose

$$1 \le x \le 1.5$$

Bisect

$$x = 1.25$$
$$y(1.25) = -0.4375$$
$$y(1)y(1.25) > 0$$

Choose

$$1.25 \le x \le 1.5$$
$$\downarrow$$
$$x = 1.38$$
$$y(1.38) = -0.0956$$
$$y(1.25)y(1.38) > 0$$

Choose

$$1.38 \le x \le 1.5$$
$$\downarrow$$
$$x = 1.44$$
$$y(1.44) = 0.0736$$
$$y(1.38)y(1.44) < 0$$

Choose

$$1.38 \le x \le 1.44$$

$$x = 1.41 \quad etc.$$

How shall we end the cycle of repetitions? Among the possibilities are:

1 $\ |x_1 - x_2| \le \varepsilon$ absolute accuracy in x

2 $\ \left| \dfrac{x_1 - x_2}{x_1} \right| \le \varepsilon$ relative accuracy in x (except for $x = 0$)

3 $\ |f(x_1) - f(x_2)| \le \varepsilon$ function values small

4   Repeat N times   a good method

Which one to use depends on the problem, but, as indicated, the author tends to favor method 4.

PROBLEMS 2.2

1   Apply the bisection method for three steps to find $\sqrt[3]{3}$, given the original interval of 1 to 2.

2   Find the zero between π and $3\pi/2$ for which $\tan x = x$. Follow the flow diagram at each step.

3   Calculate to one decimal place

$$x = \frac{1 + \sqrt{3}}{2}$$

by first finding a suitable quadratic that it satisfies.

4   Calculate to three decimal places the solution of

$$xe^x = 1$$

5   Find three decimal places at the "first" zero of $\cos x = x$.

2.3   Warnings on the bisection method

The bisection method seems to be practically "idiot proof," yet we need to be careful when we use it. First, we may have thought that the function was continuous when in fact it had a pole.

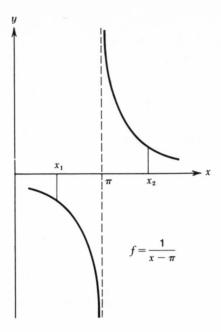

$$f = \frac{1}{x - \pi}$$

For example, if

$$f(x) = \frac{1}{x - \pi}$$

then the bisection process will come down to a small interval about $x = \pi$, and probably none of our tests will have caught the error (an overflow in evaluating the function would have been a clue). We need, therefore, to include a test for the size of $f(x_1)$ before we accept the final interval (x_1, x_2) as containing a zero.

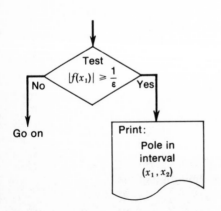

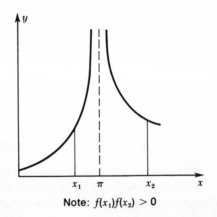

Note: $f(x_1)f(x_2) > 0$

Note that if we had used the end of repetition test $|f(x_1) - f(x_2)| < \varepsilon$, then the loop would have ended only by an overflow. Also note that double (and any even-order) poles like

$$\frac{1}{(x - \pi)^2}$$

would not be noticed at all; only odd-order poles would be searched for.

Second, if we were using one of the ε tests, it would be necessary to be careful about picking too small a value for ε. Because of the granularity of our number system, an ε which will work successfully at one zero may fail at a larger value because when we are calculating the midpoint, the granularity may force it to be one of the two original end values. Thus, we could again be caught in an endless loop. It is for this reason that the author tends to favor repeating the refinement process of bisection a fixed number of times as in test 4.

Consider test 2

$$\left| \frac{x_1 - x_2}{x_1} \right| \leq \varepsilon$$

where

$$x_1 = 0.314 \times 10^1$$
$$x_2 = 0.315 \times 10^1$$
$$\varepsilon = 10^{-4}$$

We have

$$\left| \frac{0.100 \times 10^{-1}}{0.314 \times 10^1} \right| \geq 0.300 \times 10^{-3}$$

and we do not have the fineness in spacing to achieve the inequality!

The effect of roundoff in the computation of the function values is not serious in the sense that the method does find an interval in which there is a sign change—which is all that is claimed for the method.

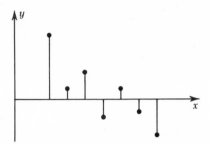

We may be misled into identifying this interval with that in which there is a mathematical zero of the function, but from the practical point of view what more can we reasonably expect if we do not also include an estimate of the roundoff noise in the function values? (This is almost never done in practice, though obviously it could be included if necessary.)

PROBLEMS 2.3

1   Draw a complete flow diagram for the heart of the bisection method with use of test 4 for the ending.
2   Show that the bisection method always finds one value when it starts with an interval with three zeros in it. Discuss the general problem of an odd number of zeros.

2.4   A search method for the bisection method

The heart of the bisection method **assumed** that there was an interval $x_1 \leq x \leq x_2$ such that

$$f(x_1)f(x_2) < 0$$

What we probably started with was the problem of finding all the real zeros in some interval $a \leq x \leq b$, and we now face the task of partially filling this gap. The solution is simple; we start with a search step of size

$$h = \frac{b-a}{n}$$

where n is the number of subintervals to be searched. Having computed $f(a)$, which we assume is not zero, we compute $a + h$ and $f(a + h)$ and then form

$$f(a)f(a+h) = \begin{cases} >0 & \text{go on} \\ =0 & \text{you have a zero at } a+h \\ <0 & \text{go to the bisection method} \end{cases}$$

For >0 and <0, we go to the next subinterval by setting

$$a = a + h$$

$$f(a) = f(a + h)$$

For = 0, we need to calculate $f(a + 2h)$, set

$$a = a + 2h$$

$$f(a) = f(a + 2h)$$

and test for zero. We repeat the process until we have done exactly n repetitions; we then stop since we are at b (within roundoff error). (But be careful of the count when = 0 occurs.)

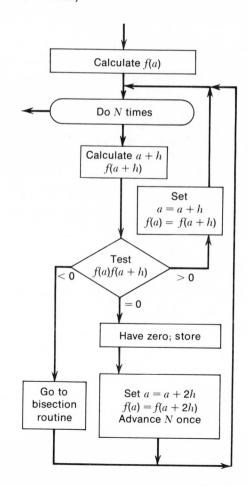

The weakness of this search method is obvious: there may be two zeros (or indeed any positive, even number of zeros) in a single subinterval and they will be missed entirely.

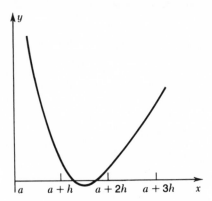

And if there are an odd number of zeros in a subinterval, then the bisection process will isolate only one of them. There appears to be no **simple** remedy for this defect (but see Sec. 2.9 for a somewhat involved solution).

The choice of the search-step size presents a typical engineering judgment situation; if the number of search steps n is very large (and hence h is very small) in order to be fairly safe, then most of the time is spent in examining intervals in which there are no zeros, which wastes a lot of computer time; but if n is very small (and therefore h is large), then the search is apt to miss some zeros.

Engineering judgment is required if

n large $\longrightarrow$ cost high
n small $\longrightarrow$ risk high

Each user must debate and settle for himself how he will choose the number of search subintervals in each particular problem.

Note that we have used n, the number of subintervals to be searched, rather than h, the step size, as the input parameter. If we used h, then we might fall into the following kind of a trap. Suppose in one instance we search from $0 \leq x \leq 5$ in steps of $h = \frac{1}{10}$ (probably $\frac{1}{8}$ or $\frac{1}{16}$ should be used on a binary machine to avoid roundoff troubles) and it works well. At a later time we are asked to examine the interval $0 \geq x \geq -5$, and without thinking that it will be necessary to change the sign of h, we try again with the result that the routine will never end.

It should be clear that the storage for the list of the zeros found must be provided, but since, at most, n intervals will have sign changes, there will be, at most, n zeros found (by this simple process). Probably the count of the number of zeros found should be given first, and a warning, as well, of any odd-order poles that were picked up.

We have deliberately left for the student the task of sketching out a complete diagram of the method (which may require some "initializing" in places).

PROBLEMS 2.4

1 Fill in the details of the complete search and bisection method.
2 What step size would you use on

$$y = \sin x - \frac{1}{x} \qquad (x > 0)$$

and why?

3 Explain why for $= 0$, we stepped two intervals forward rather than one interval.

2.5 The false-position method (a poor method)

The idea behind the false-position method (sometimes called **regula falsi**) can be found in some of our earliest mathematical records. The method is based on the observation that in trying to decrease the interval in which there is a change in the sign of the function, if one end value is large and the other is small, then the zero is probably closer to the small value than it is to the large.

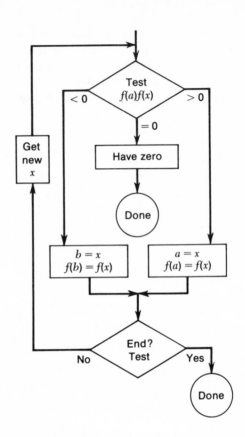

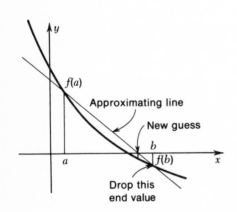

Mathematically speaking, we pass a straight line through the end points $[a, f(a)]$, and $[b, f(b)]$ and use this line as an approximation to the function $f(x)$. This line

$$y(x) = f(a) + \frac{f(b) - f(a)}{b - a} (x - a)$$

has the zero

$$x = a - \frac{f(a)(b - a)}{f(b) - f(a)} = \frac{af(b) - bf(a)}{f(b) - f(a)}$$

and this provides a next guess for the position of the zero. Note that $f(b) - f(a)$ cannot be zero but is in fact an addition of two numbers because $f(b)$ and $f(a)$ have opposite signs.

Having found the new guess, we evaluate the function at this point, namely, calculate $f(x)$, and then drop the old end point where the function value has the same sign as $f(x)$ has; thus, we keep the zero we are searching for inside the interval. If $f(x) = 0$, it is treated separately.

In the false-position method we cannot be sure of the number of steps required to decrease the interval by a preassigned amount; indeed, since the approach to the zero tends to be from one side only (see the accompanying figure), the approach can be very slow.

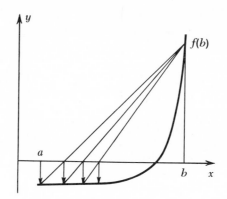

Modern computer usage often requires "time estimates" for the job to be run, and these are difficult to make for the false-position method, whereas they tend to be easy for the bisection method. For all these reasons, the author suggests avoiding the false-position method.

PROBLEMS 2.5

1 Draw a flow diagram for the complete false-position method, including the initial search for an interval.
2 Find the $\sqrt{2}$, starting with $a = 1$ and $b = 2$, by using the false-position method (three figures).
3 Find the zero of $xe^x = 1$ by the false-position method.

2.6 A modified false position (a good method)

A simple modification to the false-position method greatly improves it. The main weakness in the original method is the slow, one-sided approach to the zero. To remedy this, we arbitrarily **at each step** divide the function value that we keep by 2. The accompanying picture shows the effect of this modification.

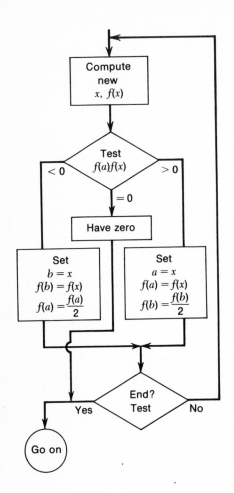

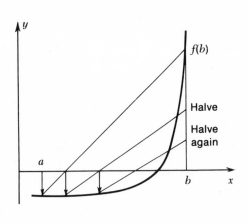

We note again that since $f(a)f(b) < 0$, there is no loss of accuracy in computing the denominator

$$f(b) - f(a)$$

since this is an addition.

This method is usually the most effective **simple** method to use so far as speed is concerned. We can no longer use the count of steps to be done as the ending test of the loop, and we have to use one of the ε tests. Thus, the timing of the process of finding a zero is no longer easily estimated.

The same search technique as used for the bisection method is available for the modified false-position method.

PROBLEMS 2.6

1 Apply the modified method to

$$y = x^2 - 2$$

2 The choice of **halving** the function value was arbitrary. Discuss other possible choices and when to use them.

3 Draw a complete flow diagram for a zero-finding routine based on the modified false-position method.

2.7   The secant method (a method to avoid)

It is often proposed that in the false-position method we always keep the two most recent values of the function and use the secant line through them as the basis for the next guess.

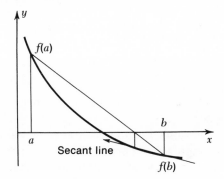

The idea that the more recent values are closer and therefore better is basically a good one, but the method can fail badly when we get both points on the same side of the zero and the approximating straight line, because of roundoff or just plain bad luck, leads us far astray.

PROBLEM 2.7

1 Make sketches showing how the secant method can fail.

2.8   Newton's method (another method to avoid)

The calculus course usually gives Newton's method for finding real zeros of a function. The idea behind the method is to fit a tangent line to the curve of the function at the point of the current estimate x_k of the zero.

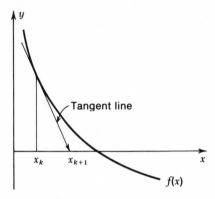

The zero x_{k+1} of this tangent line provides the next guess for the zero of the function. In mathematical notation, let x_k be the current guess. Then the tangent line is

$$y(x) = f(x_k) + f'(x_k)(x - x_k)$$

and the value where $y(x) = 0$ is

$$x_{k+1} = x_k - \frac{f(x_k)}{f'(x_k)}$$

This formula provides a method of going from one guess x_k to the next guess x_{k+1}.

Newton's method when it works is fine, but the restrictions on the method are seldom discussed. The three sketches show some of the troubles that can occur when the method is used carelessly.

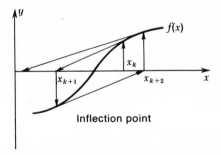

Inflection point

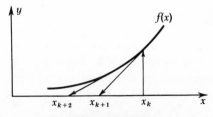

Possible multiple zero

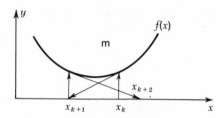

Local minimum (maximum if $f(x) < 0$)

Thus, in practice, **unless** the local structure of the function is well understood, Newton's method is to be avoided.

When Newton's method closes in on a zero, it tends on each step almost to double the number of decimal places that are accurate; thus, from two **figures** accurate we get in one step almost four figures accurate, in the next step almost eight figures, and so on. This is why, when it can be made to work, it is a good method.

Find a formula for Newton's method for the function

$$y = xe^x - 1$$

SOLUTION

$$y' = (x + 1)e^x$$

$$x_{n+1} = x_n - \frac{x_n e^{x_n} - 1}{(x_n + 1)\, e^{x_n}}$$

$$= x_n - \frac{x_n - e^{-x_n}}{x_n + 1}$$

$$= \frac{x_n^2 + e^{-x_n}}{x_n + 1}$$

Another fault of Newton's method is that it requires the differentiation of the function to find the derivative, and then the coding of the derivative, both of which certainly increase the chances for an error. But in spite of its faults we shall frequently make use of the method **when** we have information to indicate that the troubles that can plague it are not going to happen.

PROBLEMS 2.8

1 Draw a flow diagram for Newton's method.
2 Apply the method to the $\sqrt{2}$.
3 Apply the method to the $\sqrt{N}$.
4 Apply the method to the $\sqrt[3]{N}$.

5 Apply the method to the nth root of a number N.

$$\text{ANS.}\quad x_{k+1} = \frac{1}{n}\left[(n-1)x_k + \frac{N}{x_k^{\,n-1}}\right]$$

2.9 Multiple zeros

The mathematical idea of a zero of a function $y = f(x)$ is a number x_0 which makes

$$y(x_0) = f(x_0) = 0$$

In computing we must usually settle for being "close" to zero.

The idea of the **multiplicity** of a zero seems to have arisen first in connection with polynomials where a zero x_0 corresponds to the factor $x - x_0$. Since a polynomial of degree n has n factors, it is natural to say that it has n zeros even when some of the factors are repeated.

Factor theorem for polynomials

If $f(x_0) = 0$, then $x - x_0$ is a factor of $f(x)$ and conversely.

If $(x - x_0)^k$ is a factor, then x_0 is a zero of multiplicity k.

Thus, using the idea of multiplicity, we can say that a polynomial of degree n has n zeros (real or complex).

The idea of a zero can be extended to functions that can be expanded in a Taylor series about the point x_0,

$$f(x) = a_0 + a_1(x - x_0) + a_2(x - x_0)^2 + \cdots$$

The multiplicity of the zero is the number of consecutive coefficients, starting with a_0, which vanish, that is, if

$a_0 = 0$ and $a_1 \neq 0$ then a single zero

$a_0 = a_1 = 0$ and $a_2 \neq 0$ then a double zero

$a_0 = a_1 = a_2 = 0$ and $a_3 \neq 0$ then a triple zero

and so forth.

$f(x) = a_1(x - x_0) + a_2(x - x_0)^2 + \cdots$

$f(x) = a_2(x - x_0)^2 + a_3(x - x_0)^3 + \cdots$

$f(x) = a_3(x - x_0)^3 + a_4(x - x_0)^4 + \cdots$

But what are we to say about the all too common type of function

$$y = \sqrt{a^2 - x^2}$$

at $x = \pm a$?

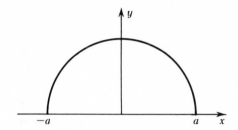

What is the multiplicity of the zero? The definition of multiplicity could be extended to such situations by defining it to be $k > 0$ if

$$\lim_{x \to x_0} \frac{f(x)}{(x - x_0)^k} = c \neq 0 \qquad \text{(nor infinite)}$$

regardless of whether k is an integer or not. When k is not an integer, the method we are going to give of examining the successive derivatives will fail.

The bisection method and some other methods as well will locate isolated zeros of odd order and will usually miss even-order zeros. If it is essential to find **all** the zeros in an interval and their multiplicities, then searching both the function and the derivative will be necessary. If the zeros of

$$y = f(x) = 0$$

are $x_1, x_2, x_3, \cdots$ (which locates all odd-order zeros) and of

$$y' = f'(x) = 0$$

are $x_1', x_2', x_3', \ldots$ (which locates all even-order zeros), then the combined set of points are the only ones that need to be examined carefully (if the zeros are all of integral order).

There will be many difficult decisions to make, however. For example, consider the two functions

$$y_1 = f_1(x) = x^2$$

and

$$y_2 = f_2(x) = x^2 + \varepsilon^2$$

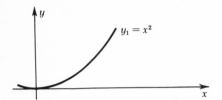

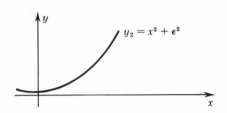

Both functions have a zero in their derivatives at $x = 0$. But how small can ε be before we decide that we shall attribute it to roundoff effects and say that both have a double zero? For the case of polynomials, the problem can be given a somewhat better answer, but for a general function, it is simply hard to decide.

2.10   Miscellaneous remarks

The problem of finding the real zeros of a function has been intensively studied for many years, and there are many methods described in the literature. We have given only a few of them. The bases for our choices have been simplicity, effectiveness, and range of application. We have tended to avoid methods whose success depends on properties of the function that the person who uses a computing machine is not likely to know and which can fail if the user cannot watch to see that the method actually works.

Frequently, the problem is not merely to find the zeros of a function

$$y = f(x)$$

but rather to find the zeros as functions of some parameter, say λ.

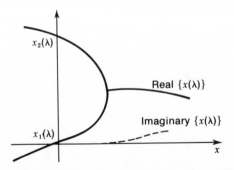

Thus, we are given

$$y = f(x, \lambda)$$

and are asked to find the zeros

$$x_i(\lambda) \qquad (i = 1, 2, \ldots, n)$$

as functions of λ. Once we have found the zeros for the first few values of λ, then we can usually use this information to guess approximately where the zeros for the next value of λ will lie—we can track the zeros as functions of λ. The methods for predicting where they might be found for the next λ value can be developed from the methods given in Chap. 8. The root-locus method for locating roots (often treated in network-design courses) can also be used to advantage at times. We shall not, however, examine the question further in this book.

COMPLEX ZEROS

3

3.1 Introduction

We now examine the problem of finding the complex zeros of an analytic function which lie in a finite region of the complex plane. By an analytic function we mean that at **all** points in the region the function can be represented by a convergent Taylor series about the point. **For convenience only**, we shall assume that we are searching a rectangular region for complex zeros.

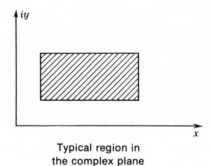

Typical region in
the complex plane

In Chap. 2 we used the notation $y = f(x)$ for the real-zero problem; we now change to the usual complex-variable notation in which the independent variable is $z = x + iy$ and the dependent variable is

$$w = f(z) = f(x + iy)$$

Once it is clearly understood how the definitions of functions in the real domain are extended to the complex domain, then it becomes obvious that $w(z)$ can be written as

$$w(z) = f(x + iy) = u(x, y) + iv(x, y)$$

where we have grouped the real and

$$
\begin{aligned}
w &= e^z - z^2 \\
&= e^{x+iy} - (x + iy)^2 \\
&= e^x(\cos y + i \sin y) - (x^2 + 2ixy - y^2) \\
&= (e^x \cos y - x^2 + y^2) + i\,(e^x \sin y - 2xy)
\end{aligned}
$$

imaginary terms separately. Thus, the single condition

$$f(x + iy) = 0$$

is equivalent to the two conditions

$$u\,(x,y) = 0$$

$$v\,(x,y) = 0$$

The $u\,(x,y)$ is called the **real part** and the $v\,(x,y)$ is called the **imaginary part** of $f(z)$.

Example   *Find the zeros of $w = \sin z$*

The basic definition is

$$\sin z = \frac{e^{iz} - e^{-iz}}{2i}$$

$$
\begin{aligned}
\sin(x + iy) &= \frac{e^{ix}e^{-y} - e^{-ix}e^{y}}{2i} \\
&= \frac{e^{-y}(\cos x + i \sin x) - e^{y}(\cos x - i \sin x)}{2i} \\
&= -\cos x\,\frac{e^{y} - e^{-y}}{2i} + \sin x\,\frac{e^{y} + e^{-y}}{2}
\end{aligned}
$$

$$v = \cos x\,\frac{e^{y} - e^{-y}}{2}$$

$$u = \sin x\,\frac{e^{y} + e^{-y}}{2}$$

Zeros require

$$u = 0$$

$$v = 0$$

From u = 0,

$$\sin x = 0$$
$$\therefore x = k\pi$$

From v = 0,

$$e^y - e^{-y} = 0$$
$$\therefore y = 0$$

Hence all zeros

$$x = k\pi$$
$$y = 0$$

The equation $u(x,y) = 0$ defines a set of curves in the complex $z = x + iy$ plane. The equation $v(x,y) = 0$ defines another set of curves, and it is **only** at the intersections of these two sets of curves (one from the $u = 0$ curves and one from the $v = 0$ curves) that $w(z) = 0$ can be realized. Thus the problem of finding the complex zeros of $w = f(z)$ is equivalent to finding the intersection of $u(x,y) = 0$ and $v(x,y) = 0$ curves, and this, algebraically, is the simultaneous solution of the two equations.

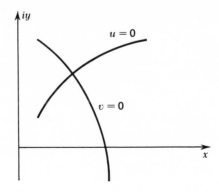

For the frequent special functions which take on only real values for real values of the argument (in spite of any i's that may be in the description of the function), then, if $z = x + iy$ is a zero, it follows that the conjugate $\bar{z} = x - iy$ is also a zero.

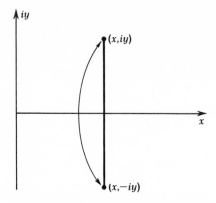

A sketch of a proof goes as follows: Since i and $-i$ are formally indistinguishable, we have at a zero

$$w = f(z) = f(x + iy) = 0$$

and

$$\overline{w} = \overline{f(z)} = \overline{f}(x - iy) = 0$$

where the long conjugate bar means that the i's in both the function and the argument are replaced by $-i$'s, and the short bar means only the i's that occur in the function itself are so replaced. By hypothesis

$$w = f(x) = u = \overline{f}(x)$$

so that

$$\overline{w} = \overline{f}(x - iy) = f(x - iy) = 0$$

Thus, in the special case of real functions of a complex variable, whenever we find a zero $x + iy$ in the upper half-plane, we know that there is the corresponding conjugate zero $x - iy$ in the lower half-plane, and we therefore need only look for the zeros that fall in the upper half-plane.

PROBLEMS   3.1

1 Find u and v if

$$\ln (x + iy) = u + iv$$

2 Write a^i as $u + iv$.

3 Write i^i as $u + iv$.

3.2 The crude method

Since the bisection method for finding real zeros is so reliable and easy to understand, it is worth extending the ideas to the problem of finding complex zeros. The bisection method has two parts, the search process (isolation) and the refinement process (improvement).

We first reexamine the search process for finding real zeros. We went step by step, using a suitably chosen step size, until we found a pair of adjacent values of $f(x)$ with one in the upper half-plane and one in the lower.

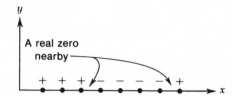

We can imagine that we had simply recorded $+$ or $-$ (or possibly 0) at each search point and then by eye isolated the real zeros. Similarly, in the crude method for complex zeros we shall search our rectangular region, using a suitably chosen grid of points in the complex $z = x + iy$ plane, and at each point we shall record the quadrant number 1, 2, 3, or 4 that

$$w = f(z)$$

falls in (we record a **0** if the function falls on the $u = 0$ or the $v = 0$ axis).

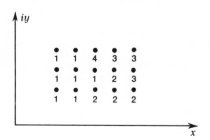

We now take colored pencils and color in the quadrants as best we can from the recorded quadrant numbers. The curves

$$u(x,y) = 0$$

divide quadrants 1 and 2, and quadrants 3 and 4, whereas

$$v(x,y) = 0$$

divide quadrants 1 and 4, and quadrants 2 and 3.

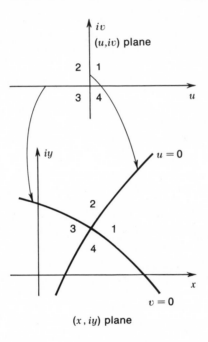

(x, iy) plane

Evidently, where (at least) four quadrants meet, we have a zero of $w = f(x)$. If eight quadrants meet at a point, then there is, as we shall see, a double zero, and so forth, but at this stage we shall confine our attention to simple zeros where four quadrants meet.

One way we can refine the accuracy of this crude method is by merely "enlarging" any small region we are interested in until we have (by successive enlargements if necessary) sufficient accuracy, or else roundoff (or the granularity of the number system) will cause trouble.

This crude method is easy to understand and is very effective, but it can be costly in machine time.

PROBLEMS 3.2

1 Prove that the x axis in a plot of quadrant numbers is a line of zeros for a real function of a complex variable.

2 Plot the quadrant numbers for the function

$$w = e^z \qquad (0 \leq x \leq 2\pi)$$

$$(0 \leq y \leq 2\pi)$$

From the plot we see why the function has no zeros.

3 Plot the quadrant numbers for $w = z^2$

3.3 An example using the crude method

As an example of how the crude method works, consider the problem of finding the complex zeros of the function

$$w = f(z) = e^z - z^2$$

which lie near the origin. The function is a real function of a complex variable so that we need only explore the upper half-plane. We try the rectangular region

$$-\pi \leq x \leq 2\pi$$

$$0 \leq y \leq 2\pi$$

The quadrant numbers are easily computed on a machine and are plotted on the figure. The x axis is a line of 0's, as it should be (there is one other almost zero number which we have marked as a 0). In drawing the curves $u = 0$ and $v = 0$, we remember that the curves in the lower half-plane are the mirror images (in the x axis) of those in the upper half but that the quadrant designations interchange 1 and 4 as well as 2 and 3.

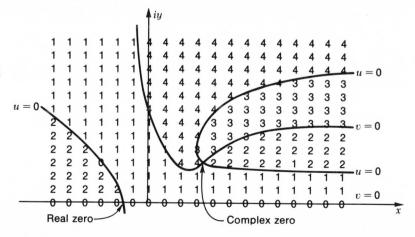

Does this picture seem to be reasonable? The real zero on the negative real axis seems to be about right because we know that as x goes from 0 through negative values, e^x decreases from 1 toward 0, whereas z^2 goes from 0 to large positive values. Thus, they must be equal at some place (and we easily see that this happens before we reach -1).

The complex zero we found is likely one of a family of zeros, the next one appearing in the band

$$2\pi \le y \le 4\pi$$

An examination of the picture shows that it is a reasonably convincing display of where the complex zero is approximately located. We could easily refine the particular region if we wished by simply placing our points in a closer mesh.

3.4   The curves $u = 0$ and $v = 0$ at a zero

At any point $z = z_0$, the Taylor expansion of a function has the form

$$f(z) = f(z_0) + f'(z_0)\frac{z - z_0}{1!} + f''(z_0)\frac{(z - z_0)^2}{2!} + f'''(z_0)\frac{(z - z_0)^3}{3!} + \cdots$$

or

$$f(z) = a_0 + a_1(z - z_0) + a_2(z - z_0)^2 + a_3(z - z_0)^3 + \cdots$$

We set for each k

$$a_k = A_k e^{i\phi_k} \qquad (A_k \text{ real})$$

and we also set

$$z - z_0 = \rho e^{i\theta}$$

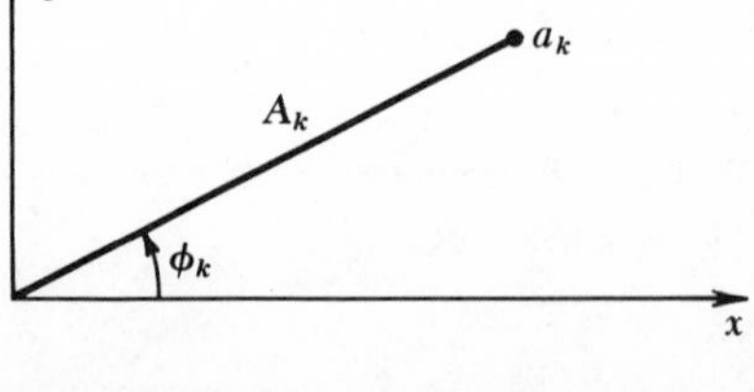

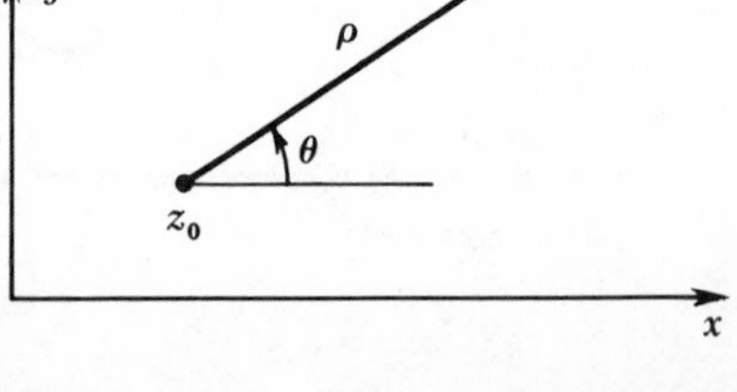

Therefore, the Taylor expansion has the form

$$f(z) = A_0 e^{i\phi_0} + A_1 \rho e^{i(\phi_1 + \theta)} + A_2 \rho^2 e^{i(\phi_2 + 2\theta)} + \cdots$$

At a simple zero $f(z_0) = 0$, then, $A_0 = 0$, and for small ρ the "immediate neighborhood of $z_0 f(z)$, looks like"

$$f(z) \approx A_1 \rho e^{i(\phi_1 + \theta)}$$

or

$$f(z) \approx A_1 \rho \left[\cos (\phi_1 + \theta) + i \sin (\phi_1 + \theta) \right]$$

The $u = 0$ curves are approximately given by

$$A_1 \rho \cos (\phi_1 + \theta) = 0$$

or

$$\theta = -\phi_1 + \frac{\pi}{2} + k\pi \qquad (k = 0, 1)$$

and the $v = 0$ curves are approximately given by

$$A_1 \rho \sin (\phi_1 + \theta) = 0$$

or

$$\theta = -\phi_1 + k\pi \qquad (k = 0, 1)$$

We see that the $u = 0$ and $v = 0$ curves intersect at right angles, and hence the quadrants we plan to color each have an angle of approximately 90° at the zero.

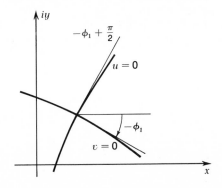

The picture we color is therefore easy to interpret at a simple zero. Note that this happened at both zeros of the example in Sec. 3.3.

At a double zero both $f(z_0)$ and $f'(z_0)$ are 0, so that the Taylor series looks like

$$f(z) = A_2 \rho^2 e^{i(\phi_2 + 2\theta)} + A_3 \rho^3 e^{i(\phi_3 + 3\theta)} + \cdots$$

We have for small ρ

$$u \approx A_2 \rho^2 \cos (\phi_2 + 2\theta)$$

$$v \approx A_2 \rho^2 \sin (\phi_2 + 2\theta)$$

$$u = 0 \longrightarrow \theta = -\frac{\phi_2}{2} + \frac{\pi}{4} + \frac{k\pi}{2}$$

$$v = 0 \longrightarrow \theta = -\frac{\phi_2}{2} + \frac{k\pi}{2}$$

and the angles of the colored quadrants are approximately 45°.

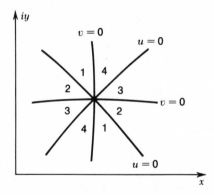

It is easy to see that for a triple zero, the colored quadrant angles will be approximately 30°, and in general for a multiplicity of m, we shall have the $u = 0$ and $v = 0$ curves meeting at approximately $\pi/2m$ radians (or 90°/m).

PROBLEMS 3.4

1 Sketch the $u = 0$ and $v = 0$ curves for

$$w = z^2 - z + \tfrac{1}{2}$$

using the lattice points

$$x = \frac{k}{4} \qquad k = 0, 1, 2, 3, 4$$

$$y = \frac{m}{4} \qquad m = 0, 1, 2, 3, 4$$

and check by using the quadratic equation formula.

2 Prove that for a zero of order m, the quadrant angles are $\pi/2m$ radians.

3.5 A pair of examples of $u = 0$ and $v = 0$ curves

The following pair of simple polynomials illustrate how the $u = 0$ and $v = 0$ curves behave at zeros and elsewhere in the plane.

The first example is a simple cubic with zeros at -1, 0, and 1. The polynomial is

$$w = f(z) = (z + 1)\, z\, (z - 1) = z^3 - z$$
$$= (x + iy)^3 - (x + iy)$$
$$= (x^3 - 3xy^2 - x) + i\,(3x^2 y - y^3 - y)$$

The real curves are defined by

$$u = x^3 - 3xy^2 - x = 0$$

or

$$x\,(x^2 - 3y^2 - 1) = 0$$

This is equivalent to two equations

$$x = 0$$
$$x^2 - 3y^2 - 1 = 0$$

The latter is an hyperbola whose asymptotes are

$$y = \pm \frac{x}{\sqrt{3}}$$

The imaginary curves are defined by

$$v = 3x^2y - y^3 - y = 0$$

This again is equivalent to two equations

$$y = 0$$

$$3x^2 - y^2 = 1$$

The latter is again a hyperbola, but this time with asymptotes

$$y = \pm \sqrt{3}\, x$$

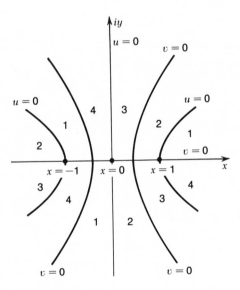

The figure we have drawn looks reasonable. In the first place, at each simple zero the real and imaginary curves cross at right angles as they should according to theory. Secondly, far out, that is going around a circle of large radius, the curves look as if they were from a triple zero, and the local effects of the exact location of the zeros tends to fade out the farther we go out.

For the second example, we move the zero from $z = -1$ to $z = 0$, which makes a double zero at $z = 0$. The polynomial is

$$w = z^2(z - 1) = z^3 - z^2$$
$$= (x + iy)^3 - (x + iy)^2$$
$$= x^3 - 3xy^2 - x^2 + y^2 + i(3x^2y - y^3 - 2xy)$$

The real curve is

$$u = x^3 - 3xy^2 - x^2 + y^2 = 0$$

Solving for y^2, we have

$$y^2 = \frac{x^2(x - 1)}{3x - 1}$$

which is easily plotted as it has a pole at $x = \frac{1}{3}$, zeros at 0 and 1, and symmetry about the x axis. As we expect, the asymptotes are parallel to

$$y = \pm \frac{x}{\sqrt{3}}$$

The imaginary curve is

$$v = 3x^2y - y^3 - 2xy = 0$$

which is

$$y(3x^2 - y^2 - 2x) = 0$$

or

$$y = 0$$

and

$$3\left(x - \tfrac{1}{3}\right)^2 - y^2 = \tfrac{1}{3}$$

The last curve has asymptotes

$$y = \pm \sqrt{3}\ \left(x - \tfrac{1}{3}\right)$$

which is what we expect.

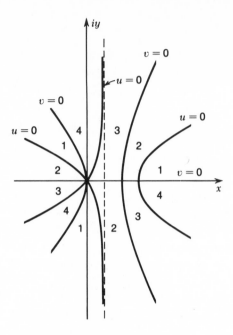

Sketching these curves, we see that far out in the complex plane they are the same as in the previous example. At the double zero, there are two real curves and two imaginary curves alternating and crossing at 45° angles. The rest of the curves look as if the real and imaginary lines were being forced together by the movement of the zero from $z = -1$ to $z = 0$, but as if they tend to repel each other strongly.

Note that in the example in Sec. 3.3, the infinite sequence of zeros must be considered in judging the reasonableness of the shapes of the curves.

3.6   General rules for the $u = 0$ and $v = 0$ curves

We cite without proof **the principle of the argument** which comes from complex-variable theory. This principle states that as you go around any contour, rectangular or not, in a counterclockwise direction, you will get a progression of quadrant numbers like

$$1, 1, 1, 2, 2, 2, 3, 3, 4, 4, 1, \ldots$$

with as many complete cycles, 1, 2, 3, 4, as there are zeros inside (we are assuming that there are no poles in the region we are searching).

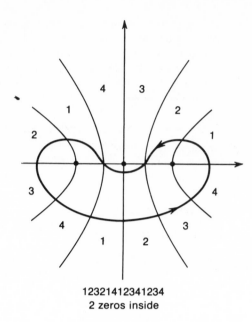

12321412341234
2 zeros inside

There may at times be retrogressions in the sequence of quadrant numbers such as

1, 1, 1, 2, 2, 3, 3, 2, 3, 3, 4, 4, 4, 1, . . .

for some suitably shaped contour, but the total number of cycles completed is exactly the number of zeros inside.

We have glossed over the instances of jumping over a quadrant number (say a 1 to a 3) and will take that up later. We have assumed that the 0's that may occur are simply neglected, as they do not influence the total number of complete cycles.

To understand this principle of the argument, the reader can try drawing various closed contours in the previous examples. No matter how involved he draws them, he will find that he will have the correct number of zeros inside when he counts +1 if he circles the zero in the counterclockwise direction, and when he counts −1 if he circles the zero in the clockwise direction. Double zeros count twice, of course.

In the general analytic function, the $u = 0$ and $v = 0$ curves can be tilted at an angle to the coordinate system, they can be somewhat distorted and involved, but they must obey the three following restraints:

1   At a zero the curves must cross alternately and be spaced according to the multiplicity of the zero.

2 Far away from any zeros, the local placement of the zeros must tend to fade out and present the pattern of an isolated multiple zero having the number of all the zeros inside (with their multiplicities).

3 The number of cycles of 1, 2, 3, 4 going counterclockwise along **any** closed contour must equal the number of zeros inside the contour (when counted properly).

These three conditions so restrict the behavior of the curves we are following as to eliminate many pathological situations and make the problem tractable.

PROBLEMS   3.6

1 Sketch the curves for the polynomial having zeros at

$$z = i, \quad z = -i, \quad z = 0$$

2 Sketch the curves for

$$w = z^4 + 1$$

3 Sketch the curves for

$$w = z^4 + 2z^2 + 1$$

3.7   An improved search method

One of the main faults of the crude method is that it wastes a great deal of machine time in calculating the function values (and corresponding quadrant numbers) at points which lie far from the $u = 0$ and $v = 0$ curves and hence give relatively little information.

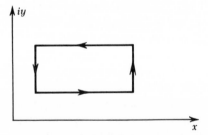

Instead of filling in the whole area of points, we propose to trace out only the $u = 0$ curves and to mark where they cross the $v = 0$ curves, which give, of course, the desired zeros.

The basic search pattern is to go counterclockwise around the area we are examining and look for a $u = 0$ curve, which will be indicated by a change from quadrant number 1 to 2 (or 2 to 1) or else from 3 to 4 (or 4 to 3).

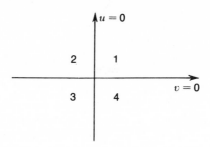

When we find such a curve, we shall track it until we meet a $v = 0$ curve, which is indicated by the appearance of a new quadrant number other than the two we were using to track the $u = 0$ curve.

Having found the general location of a zero, we shall pause to refine it, but we shall need to pass over the zero finally to continue tracking our $u = 0$ curve until it goes outside the area we are searching for zeros.

We need to know if this plan will probably locate all the zeros ("probably" depending on the step size we are using and not on the basic theory behind the plan). By the principle of the argument, the number of cycles we find is the number of zeros inside the region. Thus, the $u = 0$ (and $v = 0$) curves we care about **must** cross the boundary of the region—they cannot be confined within the region——and our search along the boundary will indeed locate all the curves we are looking for (unless the step size of the search is too large).

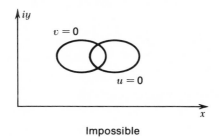

Impossible

It may happen that occasionally there is a jump in the quadrant numbers, say from 1 to 3. We can easily see that we have in one step crossed both a $u = 0$ and a $v = 0$ curve. Just where these two curves cross is, of course, not known, though probably it is near the edge.

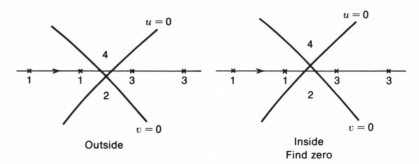

Outside

Inside
Find zero

If we want to find this zero, then we assume that it is a 1, 2, 3; whereas if we wish to ignore it, we assume that it is a 1, 4, 3. We have to modify our search plan accordingly, but this is a small detail that is not worth going into at this point.

3.8   Tracking a $u = 0$ curve

How shall we track a $u = 0$ curve? For convenience we start at the lower left-hand corner of the rectangular region we are examining for zeros and go counterclockwise, step by step, looking for a change in quadrant numbers that will indicate that we have crossed a $u = 0$ curve. When we find such a change, we construct a square† inside our region, using the search interval as one side.

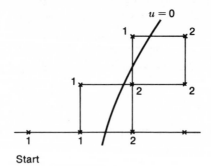

Start

The $u = 0$ curve must exit from the square so that a second side of the square will have the same change in quadrant numbers. We continue in this manner, each time erecting a square on the side that has the quadrant number change, until we find that a different quadrant number appears.

---

†Squares only if x and y are comparable in size or importance (or both); otherwise, suitably shaped rectangles to maintain relative accuracy.

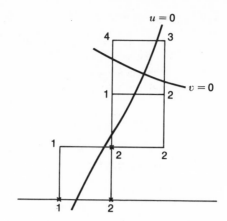

When this occurs, we know that we have crossed a $v = 0$ curve and that we are therefore near a complex zero of the function we are examining.

As a practical matter of machine economy we evaluate one point of a new square and check that side for the quadrant number change, before we evaluate the second corner of the square. In this way we occasionally save one function evaluation, but see Sec. 3.10 for why this is risky.

It is easy by eye to make the new square, but it is a bit more difficult to write the details of a program that properly chooses the two points of the next square to be examined. It is also necessary at each stage to check whether the curve we are tracking has led us out of the region we are searching,

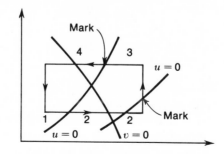

and if so, we must mark the exit so that when we come to it at a later time (while we are going around the contour), we do not track this same $u = 0$ curve again, this time in the reverse direction of course.

PROBLEMS 3.8

1 In the example of Sec. 3.3, apply the tracking method to the complex
   zero in the first quadrant.
2 Find the complex zeros as in the example in Sec. 3.3, except in the
   region $0 \le x \le 2\pi$, $2\pi \le y \le 4\pi$.

3.9 The refinement process

Having located a square during our tracking of a $u = 0$ curve which
has three distinct quadrant numbers, we have the clue that we are
near a zero. We need, therefore, to refine our search pattern and
locate the zero more accurately. The simplest way to do this is to
bisect the starting side of the square that first produced the three different quadrant numbers.

If we use this smaller size, the $u = 0$ curve crosses one of the two
halves, and selecting this half, we erect a square (of the new size). We
may or may not find a third quadrant number. If we do not, then we
continue the search with the new step size. It should be immediately
evident that within three steps we shall again have a square with three
distinct quadrant numbers.

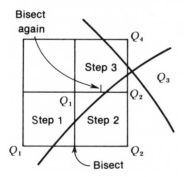

We again halve and repeat the process, continuing until we have as
small a square as we please (or run into either roundoff or the quan-
tum size of the number system of the machine).

Evidently when we stop, we know that the zero is either in the
square we have isolated or, at worst, in an adjacent square (again assuming that our step sizes are so small that the curves we are dealing
with are reasonably smooth and well behaved).

What to do when you meet a zero value while tracking one of the
curves is a small coding nuisance and is not a basic difficulty.

PROBLEM 3.9

1 Discuss how to handle a 0 value in the refinement process.

3.10 Multiple zeros in tracking

So far we have tacitly assumed that while we were tracking a $u = 0$ curve, we should come to a simple, isolated zero. But what happens when there is a double (or two very close—close for the step size we are using at the moment) or higher-order zero?

At a double zero we may find that the two quadrant numbers on the far side from the initial side of the square have the same quadrant numbers, but are reversed.

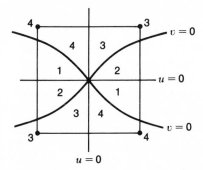

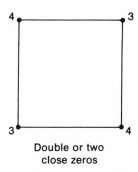

Double or two
close zeros

Thus we need, in fact, not only to check whether we have found a new quadrant number but also to check **at each step** when the quadrant number is not new that the old numbers are not diagonally opposite.

Once we sense that we are near a multiple zero, we are reasonably well off, because we can then afford relatively elaborate computing to clarify the matter. In principle it is only necessary to go around the suspected location with a sufficiently fine mesh of points to find the change in the argument and hence the number of zeros inside.

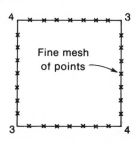

Fine mesh
of points

Once we know this, we can proceed as the situation suggests. One way is to regard it as a new problem with a much finer search size and make an "enlargement" of the region.

Another method, which somewhat simplifies the problem of multiple zeros, is not to search one curve at a time, but to go across the bottom side and find all the intervals that contain a $u = 0$ curve. Then we push the calculations up one square regardless of how many squares sidewise they go.

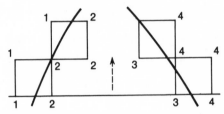

It is only when two or more $u = 0$ curves meet that we need to worry about a multiple zero. Thus, unless the second $u = 0$ curve is parallel to the bottom side of the rectangle, or so nearly so as to come to us as a total surprise, we have a very simple clue to when to suspect a multiple zero. It may be that the two are meeting head on and are the ends of a single curve, which is not hard to detect; otherwise, we are reasonably well off in making an estimate of the number of $u = 0$ curves that are coming into the local region, and hence of the multiplicity of the zero.

This method of pushing up one row at a time has the added advantage that it eliminates a lot of testing to see if the curve we are pursuing is leading us out of the region we are searching.

But let us be clear about one thing: We are vulnerable to being fooled by having chosen too large a search step, just as in the bisection method, so that the method cannot be made absolutely foolproof. What we want is a reasonably economical and safe method. All we need is a warning that there is a complicated situation at or near the location, and then we can simply apply our microscope via the enlarging process to isolate what is going on; we can repeat this enlargement process until we run into the granularity of the number system of the computing machine.

3.11 Functions of two variables

The problem of finding the real simultaneous solutions of

$$g(x,y) = 0$$

$$h(x,y) = 0$$

can be partially mapped onto the problem of finding the complex zeros of $f(x + iy) = 0$ by assigning the quadrant numbers in the obvious way:

If	Quadrant
$g > 0, h > 0$	1
$g > 0, h < 0$	2
$g < 0, h < 0$	3
$g < 0, h > 0$	4

The crude method will work, though we are not sure that the zero curves will meet at right angles at a simple zero. If we try the tracking method, we are not sure (since the principle of the argument need not apply) that we shall find all the curves as we trace the boundary. Still, the method will sometimes give useful results.

ZEROS OF POLYNOMIALS

4.1 Why treat this special method?

The problem of finding the zeros of a polynomial occurs frequently,

$$w = P_n(z)$$
$$= a_n z^n + a_{n-1} z^{n-1} + \cdots + a_0$$

and although the methods of the previous two chapters can be used, there are numerous reasons for giving a special method for finding the zeros of a polynomial. Among them are:

1. Polynomials occur frequently, and their zeros can usually be found more efficiently than can those of a general function.
2. For a polynomial we know that there are exactly n zeros to be found.

$$P_n(z) = a_n(z - z_1)(z - z_2) \cdots (z - z_n)$$

3. For most applications it is especially important to identify multiple zeros of polynomials as multiple zeros rather than as close, distinct zeros.

$$P_n(z) = a_n(z - z_1)^{k_1}(z - z_2)^{k_2} \cdots (z - z_m)^{k_m}$$

4. In the restricted case of polynomials we can give some meaning to the words "large" and "small" by suitably scaling the problem and can factor the polynomial into real linear and real quadratic factors.

Linear factors
$$z - c$$

Quadratic factors
$$z^2 - pz - q$$

5　When a zero is found, we can remove it by "deflation," that is, divide it out and get a lower-degree polynomial.

Deflation

$$\frac{P_n(z)}{z - z_1} \equiv P_{n-1}(z)$$

Furthermore, we can later reconstruct the polynomial from the zeros we have found and use this as a measure of accuracy if we wish (see Sec. 1.10).

Item 3 requires a few more words. When we ask what the zeros are to be used for, we find that if we turn up with close, distinct zeros, then for applications such as (a) finding the zeros of the characteristic equation of a linear differential equation with constant coefficients, and (b) forming the partial fractions and then doing whatever we intend to do with them (such as taking a Laplace transform), we are almost sure to run into severe roundoff problems at the later stages. Indeed, as a general remark, it is almost always true that in applications the multiple zeros can be found from the "confluent" distinct zeros and there will therefore be severe roundoff later if we have close, distinct zeros at this stage.

Example of roundoff

Solve

$$y'' + 2y' + y = 0$$
$$y(0) = 1$$
$$y'(0) = 0$$

The characteristic equation is

$$\rho^2 + 2\rho + 1 = 0$$
$$(\rho + 1)^2 = 0$$
$$\rho = -1, -1$$

The solution is therefore

$$y = e^{-x}(c_1 + c_2 x)$$
$$= e^{-x}(1 + x)$$

But if owing to roundoff the roots were

$$p_1 = -1 + \varepsilon$$
$$p_2 = -1 - \varepsilon$$

Solution

$$y = c_1 e^{-(1-\varepsilon)x} + c_2 e^{-(1+\varepsilon)x}$$

Apply initial conditions

$$1 = c_1 + c_2$$
$$0 = -(1-\varepsilon)c_1 - (1+\varepsilon)c_2$$
$$y = \frac{1}{2}\left[\left(1+\frac{1}{\varepsilon}\right)e^{-(1-\varepsilon)x} + \left(1-\frac{1}{\varepsilon}\right)e^{-(1+\varepsilon)x}\right]$$

Note roundoff trouble, especially near $x = 0$.

It is for this reason that the proposed routine differs so significantly from those usually offered by the pure mathematicians who are usually anxious to show how well they can separate zeros; we are equally anxious to find all multiple zeros so that at a later stage we shall not face roundoff disaster. Different objectives produce different methods.

We shall confine the treatment to real polynomials; thus, we propose to find the real linear and real quadratic factors. If the complex zeros are wanted, they are easily found from the real quadratic factors by using the quadratic formula for the zeros.

Another purpose of the chapter is to extend our analysis of Newton's method.

4.2 Scaling

The idea of scaling in this case is to make a transformation on the polynomial that results in another polynomial of the same degree n whose zeros are simply related to those of the original equation. We have a number of such transformations available.

The simplest transformation is to divide the whole polynomial by the leading coefficient, which makes the first coefficient equal to 1.

$$\frac{1}{a_n} P_n(z) = z^n + \frac{a_{n-1}}{a_n} z^{n-1} + \cdots + \frac{a_o}{a_n} = 0$$

Another transformation is to replace z by a new variable z'

$$z = \frac{z'}{b}$$

or

$$z' = bz$$

and multiply the equation by b^n to produce a new polynomial for which the coefficient a_{n-k} is multiplied by b^k.

$$b^n P_n\left(\frac{z}{b}\right) = b^n\left[\left(\frac{z}{b}\right)^n + \frac{a_{n-1}}{a_n}\left(\frac{z}{b}\right)^{n-1} + \cdots + \left(\frac{a_0}{a_n}\right)\right] = 0$$

becomes

$$= z^n + \left(\frac{a_{n-1}b}{a_n}\right)z^{n-1} + \cdots + \left(\frac{a_0 b^n}{a_n}\right) = 0$$

If we now pick b to make the coefficient of the constant term equal to 1 in size, then we have made a second transformation toward a form that is more easily understood.

Set

$$\frac{a_0 b^n}{a_n} = 1$$

or

$$b = \sqrt[n]{\left|\frac{a_n}{a_0}\right|}$$

$$\frac{1}{a_n} P_n(-z) = (-1)^n\left[z^n - \frac{a_{n-1}}{a_0}z^{n-1} + \cdots + (-1)^n\frac{a_0}{a_n}\right]$$

A third transformation is the substitution $z = -z'$ which changes the sign of alternate coefficients. It could be used to confine our search to the positive half of the plane if we wished, but this seems, in practice, to give us no advantage.

A fourth transformation is the substitution

$$z = \frac{1}{z'}$$

followed by multiplying the whole equation by $(z')^n$, which has the effect of reversing the order of the coefficients and of taking the zeros that lie outside the circle $|z| = 1$ in the complex plane and bringing them inside.

$$(z')^n P_n\left(\frac{1}{z'}\right) = (z')^n \frac{a_n}{(z')^n} + \frac{a_{n-1}}{(z')^{n-1}} + \cdots + a_0$$
$$= a_0(z')^n + a_1(z')^{n-1} + \cdots + a_n$$

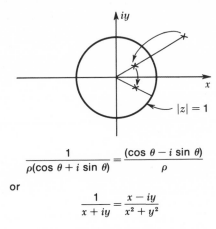

$$\frac{1}{\rho(\cos\theta + i\sin\theta)} = \frac{(\cos\theta - i\sin\theta)}{\rho}$$

or

$$\frac{1}{x+iy} = \frac{x-iy}{x^2+y^2}$$

As a result, we can search the two finite intervals $-1 \le z \le 1$ and $-1 \le z' \le 1$ when we are considering the real zeros, and the two circular regions $|z| \le 1$ and $|z'| \le 1$ when we are considering complex zeros.

The first two transformations may be used to make the first and last coefficients equal to 1 in size; this has the effect of scaling the size of the numbers we are going to handle. If we do this, then we know that the product of all the zeros is $(-1)^n$.

$$(z-z_1)(z-z_2)\cdots(z-z_n) = z^n - (z_1 + z_2 + \cdots + z_n)z^{n-1} +$$
$$\cdots + (-1)^n(z_1 z_2 \cdots z_n) = 0$$

Therefore,

$$|z_1 z_2 \cdots z_n| = a_0$$

Therefore, if there are some zeros of large absolute value, then there must also be compensating small ones. By searching the two finite regions $|z| \le 1$ and $|z'| \le 1$, we are tending to favor the risk of underflow rather than overflow, which, as we observed before, is preferable.

In actual practice, the scaling we have talked about would probably produce roundoff effects in the original problem before we even start to find the zeros. It is sometimes proposed to limit the scaling to the nearest powers of 2 (on a binary machine) since there is no roundoff effect in the exponents. We prefer to do a more practical thing, not actually to do the scaling but to remember what the scale factors would have been, and then use these factors when we come to make the various choices when something is "big" or "small" or what to do next. As a result, it will not be the unit circle that we use, but rather a circle of a convenient radius approximately equal to geometric mean of the absolute values of the zeros, namely,

$$\rho = \sqrt[n]{\left|\frac{a_n}{a_0}\right|}$$

PROBLEMS   4.2

1 Scale

$$2z^2 - 7z + 9 = 0$$

2 If you did decide to scale the equation, write the expression for the appropriate power of 2 value of ρ.

3 Scale

$$3z^4 - 5z^3 + 7 = 0$$

4.3   The search for the real zeros

Our first problem is to try to find the real zeros and divide out the corresponding factors so that we shall be left with the problem of finding the real quadratic factors of an even-degree polynomial. Since real zeros are so much easier to find than complex zeros, it is well to do this first, but unfortunately, because we do not know exactly how many real zeros there will be, we are not in a position to say when we have found them all. Thus as a matter of practice, we shall make a reasonable effort to find all the real zeros and on occasion shall later find that a quadratic factor has a pair of real zeros. There will be, therefore, a small inconsistency in this matter because we shall later be assuming that the quadratic factors have no real zeros when we reason how the process will go. The difference is apparently not serious in practice.

We know that the polynomial has exactly n zeros. How fine a search step size for the bisection method (or modified false position) shall we use when we search the intervals in z and z'? The finer the search, the more probably we shall uncover the location of possible odd-order zeros. If at the same time we also examine the derivative of the polynomial, we shall have the possible locations of the even-order zeros. As a practical compromise, we shall explore $-\rho \leq z \leq \rho$ in $4n$ steps of equal size, and $-1/\rho \leq z' \leq 1/\rho$ in $4n$ steps of equal size. When we find a sign change, we can decrease the interval, using either the bisection method or the modified false-position method, until we have the accuracy we wish or run into roundoff evaluation troubles.

In order to handle the real zeros and their multiplicities, we write the polynomial in the Taylor series form†

$$P_n(z) = \overline{a}_0 + \overline{a}_1(z - z_k) + \overline{a}_2(z - z_k)^2 + \cdots + \overline{a}_n(z - z_k)^n$$

where z_k is a possible real zero. To find the coefficients $\overline{a}_j$ we proceed as we did in finding the binary representation of an integer (which is simply the representation of the integer as a sum of increasing powers of 2) and divide the polynomial by $z - z_k$. The remainder is $\overline{a}_0$. We next divide the quotient by $z - z_k$, which gives us the remainder $\overline{a}_1$, and so forth.

If

$$P_n(z) = z^4 - 4z^3 + 7z^2 - 4z + 3$$
$$zk = 1$$

then

$$
\begin{array}{r|rrrrr}
1 & 1 & -4 & 7 & -9 & 3 \\
  &   & 1 & -3 & 4 & -5 \\
\hline
1 & 1 & -3 & 4 & -5 & \boxed{-2} \leftarrow \overline{a}_0 \\
  &   & 1 & -2 & 2 & \\
\hline
1 & 1 & -2 & 2 & \boxed{-3} \leftarrow \overline{a}_1 \\
  &   & 1 & -1 & \\
\hline
1 & 1 & -1 & \boxed{1} \leftarrow \overline{a}_2 \\
  &   & 1 & \\
\hline
  & 1 & 0 \leftarrow \overline{a}_3 \\
  & \uparrow \\
  & \overline{a}_4
\end{array}
$$

or

$$P_n(z) = (z - 1)^4 + (z - 1)^2 - 3(z - 1) - 2$$

---

†The overbar does **not** in this case mean the complex conjugate.

We can regard this form of the polynomial in three ways:

1   As an expansion in powers of $z - z_k$
2   As the Taylor expansion of the polynomial about $z = z_k$ where

$$\overline{a}_m = \frac{1}{m!} \frac{d^m P_n}{dz^m}$$

3   As one stage in the classical Horner's method

We will adopt the first view.

Evidently the number of consecutive zero coefficients, starting with $\overline{a}_0$ gives the multiplicity of the zero, and the z_k which make the coefficient $\overline{a}_1$ zero are the zeros of the first derivative. Thus, in our search process we need to follow up on sign changes in both the function value $\overline{a}_0$ and the derivative value $\overline{a}_1$, until:

1   The zero of the function is located sufficiently accurately.
2   A zero in the derivative either yields a zero in the function at the same place, or else it is clearly not a multiple zero.

4.4   Roundoff

From the zeros of the polynomial and its derivative we are in a position to estimate their multiplicities. For this we need some criterion for deciding when various coefficients in the expansion are zero. We will arbitrarily adopt the rule that if a number is less than the roundoff that could have come from the roundoff in the initial coefficients of the polynomial (computed in a certain way),

---

Consider

$$P(z) = z^2$$

and

$$P(z) = z^2 + \varepsilon^2$$

Both have $P'(0) = 0$. How small must ε be to decide $P(0) = 0$ has a double zero at $z = 0$ in the second case?

---

then we may call the number zero is we please, and we shall use this freedom to try to find multiple zeros as often as possible (see item 3 of Sec. 4.1).

We will take the roundoff in the original coefficients in the floating point form

$$a_m(1 + \varepsilon_m)$$

where for convenience we assume

$$|\varepsilon_m| \leq 2^{-(N+1)} \qquad N = \text{mantissa length}$$

(any other bound is easily done).

In the division process that we used to get the expansion in powers of a factor $z - z_k$, where z_k is a suspected zero, we can find the maximum roundoff that occurs and is due **only** to the roundoff in the original coefficients. This can be done by carrying out a similar division process but using the absolute values of the coefficients and the absolute value of the suspected value z_k. Thus, in the division process for the roundoff estimation there will always be an addition at each stage, and some suitable choice of roundoff error in the a_m can be found to lead to this addition.

$$
\begin{array}{c|cccc}
|z_k| & |a_n| & |a_{n-1}| & \cdots & |a_0| \\
 & |a_n z_a| & & & \\
\hline
|z_k| & |a_n| & \cdots & & |\bar{r}_0| \\
 & & \cdots & & \\
\hline
 & |a_n| & \cdots & &
\end{array}
$$

As a result, if we carry out the process on the absolute values in parallel with the original division process, and at the end simply multiply the remainders by $2^{-(N+1)}$, we will have estimates of the maximum roundoff of the numbers in the corresponding positions to those in the original division process.† Whenever this calculated bound exceeds the corresponding remainder in size, we may call the remainder "zero" if we please—ignoring any possible conflicts that might arise from incompatible requirements on the roundoff assumptions needed to make different numbers zero.

4.5 Multiple real-zero location and estimation

We now assemble the elements for making the estimates of both the real-zero locations and their corresponding multiplicities.

---

†Assuming that the evaluation is done double precision; if single precision, then add 1's in the proper places.

We are going to search the interval $0 \le z \le \rho$ in $2n$ steps (where n is the degree of the polynomial) and then do the same for $0 \ge z \ge -\rho$, $0 \le z' \le 1/\rho$, and $0 \ge z' \ge -1/\rho$. Each time we find a sign change in the values of the polynomial, we know that there is an odd number of zeros in the intervals. The corresponding sign change in the derivative gives us the possibilities for even-order multiplicities.

When we find an interval to search, we use the bisection method, or else the modified false-position, to refine the zero location. Having found the zero z_k reasonably accurately, we resort to the division process (with its accompanying roundoff estimation process) to attack the multiplicities. For each multiplicity we can use the first non-vanishing (in the roundoff sense) remainder to make the immediately preceding remainder as small as possible. In the case of a double zero, this process amounts to taking the local maximum or minimum as the best estimate of the pair of zeros that we are identifying as a double zero; for three close zeros we are taking the local inflection point; and so forth.

When we find the improved value as accurately as we can, we need to check that the roundoff bounds have not been exceeded. If they have been, we must revert to searching for a lower-order multiplicity; if they have not been exceeded, then we **deflate** the polynomial by removing that factor with the proper multiplicity and take the resulting quotient as the deflated polynomial.

Given a double zero

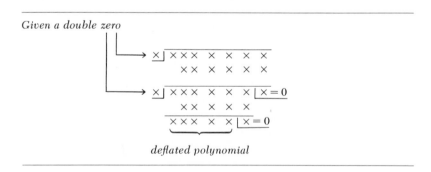

deflated polynomial

We want to remove the zeros in increasing order of size to minimize the propagated roundoff effects, so that for each interval we search in turn the positive zeros, the negative zeros, and then the reciprocal zeros, both positive and negative.

We might think that we should do a final refinement of the zeros by using the original polynomial, but many users seem to prefer having a **consistent** set of zeros that **as a set** comes close to reconstructing the original polynomial, rather than having a set of perhaps individually more accurate zeros, but as a set probably somewhat more incon-

sistent in that they have independent small errors rather than somewhat correlated errors.

PROBLEM 4.5

1 Show that the search and deflation process results in a polynomial of
   even degree (allow for roundoff).

4.6 Complex zeros

Having found and removed all the real zeros, or perhaps having by an
unfortunate chance missed one or more pairs of real zeros, we now
have an even-degree polynomial to search for the quadratic factors
and their corresponding multiplicities. Only if the degree is four or
greater is there anything to be done.

We shall assume our quadratic factors in the form

$$Q(x) = z^2 - pz - q$$

(the reason for the minus signs is in analogy with the linear factor situation).

By repeated division of the polynomial by the trial quadratic factor
we can write the polynomial as an expansion in powers of $Q(x)$ instead of in powers of $(x - x_k)$, as was discussed in the case of real
roots in Sec. 4.3.

Modified synthetic division for quadratic factors

$$
\begin{array}{c}
\times\times\cdots\times\times \\
1-p-q\ \overline{)\times\times\times\times\cdots\times\times} \\
\times\times\times\times\times \\
\times\times\ldots\times \\
\overline{\times\times\times\times\ldots.\lfloor\times\times}
\end{array}
$$

or

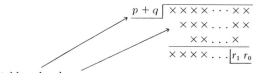

$$
\begin{array}{c}
p+q\ \overline{\lceil\times\times\times\times\cdots\times\times} \\
\times\times\times\ldots\times\times \\
\times\times\ldots\times \\
\overline{\times\times\times\times\ldots.\lfloor r_1\ r_0}
\end{array}
$$

Add with + here.

We have $Q(x)$ as a factor if, and only if, the two coefficients of the remainder both vanish. There is a double factor if this is true for the first two remainders, and so forth. Thus, we have a situation exactly analogous to that of real factors. But here we have two parameters p and q to adjust to make two numbers, the linear and the constant terms of the remainder, vanish.

In more detail, given

$$P_{2n}(z) = a_{2n}z^{2n} + a_{2n-1}z^{2n-1} + \cdots + a_0$$

divide $P_{2n}(z)$ by $Q(z)$ to get a remainder $r_1 z + r_0$; thus,

$$P_{2n}(z) = Q(z)P_{2n-2}(z) + r_1 z + r_0$$

Now, divide $P_{2n-2}(z)$ by $Q(z)$ to get

$$P_{2n-2}(z) = Q(z)P_{2n-4}(z) + r_3 z + r_2$$

and so forth, and assemble the results

$$P_{2n}(z) = a_{2n}Q^n(z) + (r_{2n-1}x + r_{2n-2})Q^{n-1}(z) + \cdots + (r_5 z + r_4)Q^2(z) + \\ (r_3 z + r_2)Q(z) + (r_1 z + r_0)$$

which is our expansion of $P_{2n}(z)$ in powers of $Q(z)$.

For example

$$P_4(z) = z^4 + 2z^3 + 2z + 15$$
$$Q(z) = z^2 - 2z + 3$$

2	−3	1	2	0	2	15
			2	−3	−12	−15
				8	10	
		1	4	5	0	0

Quotient　　*Remainder*

$$z^2 + 4z + 5$$

PROBLEM   4.6

1　Write $z^4 + 2z^3 + 2z + 15$ in powers of $x^2 + x + 2$.

4.7 Bairstow's method

Since the coefficients of the remainder

$$r_1 = r_1(p,q)$$
$$r_0 = r_0(p,q)$$

are analytic functions in p and q, we may expand them in a Taylor series about the current point (p,q). The value at any nearby point (p^*,q^*) in the (p,q) plane is given by

$$r_1(p^*,q^*) = r_1(p,q) + \frac{\partial r_1}{\partial p}\Delta p + \frac{\partial r_1}{\partial q}\Delta q + \cdots$$

$$r_0(p^*,q^*) = r_0(p,q) + \frac{\partial r_0}{\partial p}\Delta p + \frac{\partial r_0}{\partial q}\Delta q + \cdots$$

where

$$\Delta p = p^* - p$$
$$\Delta q = q^* - q$$

As a first approximation, we drop all the terms beyond the linear ones. Next we pick (p^*,q^*) so that

$$r_1(p^*,q^*) = 0$$
$$r_0(p^*,q^*) = 0$$

gives us the improvements in our guess at p and q. This is essentially Newton's method in two variables.

To get the required partial derivative, we observe that $P_{2n}(z)$ does not depend on p or q, and we differentiate

$$\frac{\partial P_{2n}}{\partial p} \equiv \{\cdots\}\, Q(z) + (r_3 z + r_2)\frac{\partial Q}{\partial p} + \frac{\partial r_1}{\partial p}z + \frac{\partial r_0}{\partial p} = 0$$

$$\frac{\partial P_{2n}}{\partial q} \equiv \{\cdots\}\, Q(z) + (r_3 z + r_2)\frac{\partial Q}{\partial q} + \frac{\partial r_1}{\partial q}z + \frac{\partial r_0}{\partial q} = 0$$

These equations are **identities** in z, hence true for all values of z, and in particular for those which are zeros of $Q(z)$. Using one of these, we have

$$(r_3 z + r_2)(-z) + \frac{\partial r_1}{\partial p} z + \frac{\partial r_0}{\partial p} = 0$$

$$(r_3 z + r_2)(-1) + \frac{\partial r_1}{\partial q} z + \frac{\partial r_0}{\partial q} = 0$$

But since $z^2 = pz + q$ and we are operating formally, we regard z as a complex number and equate the linear and constant terms separately to zero.† Thus, we get the four equations

$$r_3 p + r_2 = \frac{\partial r_1}{\partial p}$$

$$r_3 q = \frac{\partial r_0}{\partial p}$$

$$r_3 = \frac{\partial r_1}{\partial q}$$

$$r_2 = \frac{\partial r_0}{\partial q}$$

The Newton approximations from the truncated Taylor expansions become

$$(r_3 p + r_2)\, \Delta p + r_3 \, \Delta q = -r_1$$

$$(r_3 q)\, \Delta p + r_2 \, \Delta q = -r_0$$

which are the Bairstow equations for the change to make in our approximations to p and q.

The determinant of these equations is

$$D = r_2{}^2 + r_2 r_3 p - r_3{}^2 q$$

$$= r_3{}^2 \left\{ \left(\frac{r_2}{r_3} \right)^2 + \left(\frac{r_2}{r_3} \right) p - q \right\}$$

---

†We are using the fact that if $az + b = 0$, with a and b real and z complex, then $a = b = 0$.

From the last form (valid only if $r_3 \neq 0$), we see that the ratio r_2/r_3 is a zero of $z^2 + pz - q = 0$ (note + sign), and if the original quadratic $Q(x) = z^2 - pz - q$ has complex zeros, then $D \neq 0$. If $r_2 = r_3 = 0$, then clearly we are "near" a possible multiple zero. The possibility that $q = 0$ and $r_2 = 0$ is also unlikely, for although we shall use $q = 0$ as our starting approximation, we do not expect to have $q = 0$ as a final approximation since this implies that $z = 0$ (or $z = \infty$) is a zero and we removed these before we started this part of the search.

4.8   Revisions of Newton's method

The weakness of Newton's method is that it can wander and fail to converge properly, and one method of coping with this is to introduce a "distance function" to measure how far the remainder is from zero.

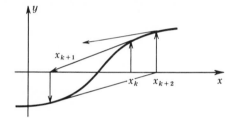

The problem is what distance function to use. One reasonable one is

$$|P_{2n}(z)|$$

evaluated at a zero of $Q(z) = z^2 - pz - q = 0$. We will use

$$|P_{2n}(z)|^2 = (r_1 z + r_0)(r_1 \bar{z} + r_0)$$
$$= r_1^2(-q) + r_1 r_0 p + r_0^2$$

and evaluate this distance function at the end of each step (which is really the beginning of the next step). If the distance has not been decreased by the step, then we will back up and divide both Δp and Δq by 2 and try the smaller step. Repeating this enough times we shall finally come to an acceptable step in the indicated direction.

Use

$$\frac{\Delta p}{2} \qquad \frac{\Delta q}{2}$$

if $|P_{2n}|$ does not decrease, and try again.

When using this general trick, it is wise to add a second, namely, that for the next step the Δp and Δq should not be allowed to increase by more than a factor of 2 over that used on the last step. This greatly improves slow convergence.

Experience shows that these two tricks will make the method converge even in the presence of multiple factors.

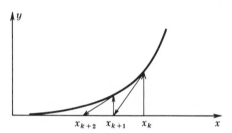

We must pause and examine why we expect this application of Newton's method, with these modifications, to work better than in the case of a single real variable. How, in particular, are we to avoid local minima?

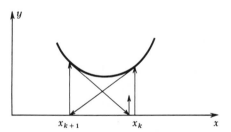

In the theory of a complex variable there is a minimum modulus theorem (much like the maximum modulus theorem), which says that if at a point z_k in the z plane, $f(z_k) \neq 0$ [and $f(z)$ is not a constant], then there are values of $f(z)$ in the immediate neighborhood such that

$$|f(z)| < |f(z_k)|$$

We are, of course, not operating in the z plane but are in the p, q plane, where

$$Q(z) = z^2 - pz - q = 0$$

gives the relationship between points in the two planes.

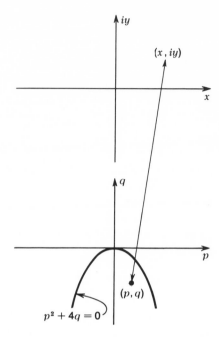

Provided we stay away from the curve of multiple (real) zeros of the quadratic factor $Q(z) = 0$, namely,

$$p^2 + 4q = 0$$

then locally this gives a one-to-one mapping of points in one plane into points in the other. This in turn means that if the point (p,q) does not make $|P_{2n}|^2 = 0$, then there are nearby points which will reduce the value of the distance function.

Even if there are points in the immediate neighborhood which reduce the value of $|P_{2n}|^2$, it is not obvious that we shall be pointed in such a direction by the Δp and Δq obtained from the Bairstow equations. Indeed, it is not always true that we shall be pointed in a favorable direction; in the case of a polynomial having no odd-degree terms, that is, a polynomial with only even powers of z,

---

$$z^6 + 3z^4 - 2z^2 + 7 = 0$$

For even-degree terms only.

---

then starting our process with the usual values

$$p = 0 \qquad q = 0$$

We find

$$r_1 = r_3 = 0$$

The Bairstow equations become

$$r_2 \, \Delta p = 0$$

$$r_2 \, \Delta q = -r_0$$

which will leave $p = 0$, and this in turn assures us that r_1 and r_3 will also both stay zero. And this in turn can keep us from finding a factor of the polynomial.

The solution of this trouble is easy. If the polynomial has only even-degree terms, then the substitution of $z^2 \longrightarrow z'$ will produce a polynomial of half the degree.

The polynomial

$$z^6 + 3z^4 - 2z^2 + 7 = 0$$

$$z^2 \longrightarrow z'$$

becomes

$$(z')^3 + 3(z')^2 - 2z' + 7 = 0$$

If in turn this has only even-degree terms, then we repeat the transformation until we finally come to a poylnomial which has at least one term of odd degree. The presence of this term will assure us that except in a fantastically unlikely combination of roundoffs we shall get a nonzero value for Δp. In loose language, we might have been exactly on a ridge of a saddle of the function $P_{2n}(z)$ and have needed a little push to find our way down to lower values; the term of odd degree will supply such a push.

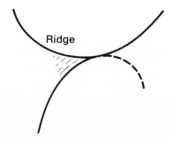

Ridge

When shall we apply the test for even-degree terms only? One place would be immediately after we have checked that both $a_0 \neq 0$ and $a_N \neq 0$. Another place would be when we are about to start the search for quadratic factors. To be safe, we can try it both places.

Note that the reduction can produce a polynomial of odd degree which would have at least one real zero; hence, we go back and search for real zeros again, even for multiple zeros!

We have, now, to consider how to undo the effects of the transformations $z^2 \longrightarrow z'$. Probably the simplest way is to take the zeros of the factors we have found, write them in polar coordinate form, and apply de Moivre's theorem, using the proper multiplicities.

De Moivre's theorem

$$(\cos \theta + i \sin \theta)^n = \cos n\theta + i \sin n\theta$$

Alternatively, we may note that it is easy to factor

$$z^4 - pz^2 - q$$

into

$$(z^2 + az + \sqrt{-q})(z^2 - az + \sqrt{-q})$$

where

$$a^2 = 2\sqrt{-q} + p$$

We have to repeat this process the proper number of times (plus the simpler process for any real zeros that may appear).

4.9 A minimization approach

We are still not certain that Bairstow's equations will lead us to a smaller value—only reasonably confident that the modifications have removed the worst objections to what is basically Newton's method in two variables. If we insist on being mathematically sure that we are pointed in a direction in which we can find a smaller value, then we need to use some of the minimizing techniques of Chap. 9.

We are trying to minimize the distance

$$f(p,q) \equiv |P_{2n}(z)|^2 = -qr_1{}^2 + pr_1r_0 + r_0{}^2.$$

Consider "a level curve" along which

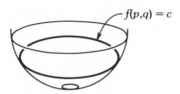

$f(p,q) = c$

$f(p,q)$ is a constant,

$$f(p,q) = c$$

A nearby point on the same level curve would be

$$f(p + \Delta p, \, q + \Delta q) = c$$

Their difference is

$$f(p + \Delta p, \, q + \Delta q) - f(p,q) = 0 = \frac{\partial f}{\partial p} \, \Delta p + \frac{\partial f}{\partial q} \, \Delta q +$$

$\cdots$ higher-order terms

The tangent line at the point (p,q) is given by the linear terms

$$\frac{\partial f}{\partial p} \, \Delta p + \frac{\partial f}{\partial q} \, \Delta q = 0$$

where now the point $(p + \Delta p, \, q + \Delta q)$ is on the tangent line.

Perpendicular lines have negative reciprocal slopes,

$$m_1 = -\frac{1}{m_2}$$

A line perpendicular to this level line is

$$\frac{\partial f}{\partial q} \, \Delta p - \frac{\partial f}{\partial p} \, \Delta q = 0$$

and points in the direction of "steepest descent."

How far do we want to go in this direction? We know in this case that we have a minimum of zero. Expanding $f(p,q)$ about the point (p_0,q_0) on the level curve, we have, keeping only the linear terms,

$$f(p,q) = f(p_0,q_0) + \frac{\partial f}{\partial p}\,\Delta p + \frac{\partial f}{\partial q}\,\Delta q$$

We want $f(p,q) = 0$ so that this equation plus the equation for the direction,

$$\frac{\partial f}{\partial p}\,\Delta p + \frac{\partial f}{\partial q}\,\Delta q = -f(p_0,q_0)$$

$$\frac{\partial f}{\partial q}\,\Delta p - \frac{\partial f}{\partial p}\,\Delta q = 0$$

[where the derivatives are evaluated at the point (p_0,q_0)], determines the step sizes Δp and Δq

$$\Delta p = \frac{-f\,(\partial f/\partial p)}{(\partial f/\partial p)^2 + (\partial f/\partial q)^2}$$

$$\Delta q = \frac{-f\,(\partial f/\partial q)}{(\partial f/\partial p)^2 + (\partial f/\partial q)^2}$$

We now find the partial derivatives

$$f(p,q) = -qr_1{}^2 + pr_1r_0 + r_0{}^2$$

$$\frac{\partial f}{\partial p} = -pqr_1r_3 - 2r_1r_2q + p^2r_0r_3 + pr_0r_2 + 2r_0r_3q$$

$$\frac{\partial f}{\partial q} = -r_1{}^2 - 2r_1r_3q + pr_1r_2 + pr_0r_3 + 2r_0r_2$$

where we have used the earlier relations (Sec. 4.7)

$$\frac{\partial r_1}{\partial p} = r_3 p + r_2$$

$$\frac{\partial r_0}{\partial p} = r_3 q$$

$$\frac{\partial r_1}{\partial q} = r_3$$

$$\frac{\partial r_0}{\partial q} = r_2$$

The trouble that will appear is that at the minimum

$$\frac{\partial f}{\partial p} = 0 \qquad \frac{\partial f}{\partial q} = 0$$

so that the denominator approaches zero quadratically—but so does the numerator, so that in principle the limit is correct. Underflow is, however, a problem.

4.10 Multiple factors

The problem of multiple factors is theoretically quite difficult, as shown by the following example. In practice it is not so bad; rarely do more than double zeros occur in actual problems.

$$f_1(z) = (z+1)^8$$
$$= z^8 + 8z^7 + \cdots + 70z^4 + \cdots + 8z + 1$$
$$f_2(z) = (z+1)^8 + (\tfrac{1}{10})^8$$
$$= z^8 + 8z^7 + \cdots + 70z^4 + \cdots + 8z + 1 + (\tfrac{1}{10})^8$$

To eight decimal places the coefficients of $f_1(z)$ are the same as those of $f_2(z)$. But the zeros of the first are at $z = -1$, whereas those of the second are on a circle of radius $\tfrac{1}{10}$ about $z = -1$.

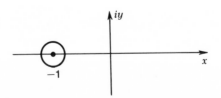

The theory of real zeros was carefully developed so that the obvious parallel would apply for quadratic factors with complex zeros, and so nothing more need be said. The same form for bounds on the round-off are used, except, of course, we divide by the trial quadratic at each step.

SIMULTANEOUS LINEAR EQUATIONS AND MATRICES

5.1 The idea of Gaussian elimination

The problem of finding the numerical solution of a set of simultaneous linear equations

$$a_{1,1}x_1 + a_{1,2}x_2 + \cdots + a_{1,n}x_n = b_1$$

$$a_{2,1}x_1 + a_{2,2}x_2 + \cdots + a_{2,n}x_n = b_2$$

$$\cdots\cdots\cdots\cdots\cdots\cdots\cdots\cdots\cdots\cdots\cdots\cdots\cdots\cdots\cdots$$

$$a_{n,1}x_1 + a_{n,2}x_2 + \cdots + a_{n,n}x_n = b_n$$

occurs frequently. It is necessary to examine the theoretical as well as the practical aspects of the problem in order to understand what happens in actual practice.

With the **sigma notation,**

The sigma notation

$$\sum_{i=1}^{n} x_i = x_1 + x_2 + \cdots + x_n$$

Special cases:

$$\sum_{i=1}^{n} c = c \sum_{i=1}^{n} 1 = cn$$

$$\sum_{i=1}^{1} x_i = x_1$$

$$\sum_{i=1}^{0} x_i = 0$$

Notice in

$$\sum_{i=1}^{n} x_i = \sum_{j=1}^{n} x_j$$

i and j are "dummy indices."

---

the system can be written as

$$\sum_{j=1}^{n} a_{1,j}x_j = b_1$$

$$\sum_{j=1}^{n} a_{2,j}x_j = b_2$$

. . . .

$$\sum_{j=1}^{n} a_{n,j}x_j = b_n$$

or even more simply as

$$\sum_{j=1}^{n} a_{i,j}x_j = b_i \qquad i = 1, 2, \ldots, n$$

This compact notation is useful but should fool no one into thinking that anything fundamentally new has appeared.

An example of the importance of notation

Consider multiplying CCLXIV *by* DXLIX *or multiplying* 264 *by* 549.

Which is easier?

The Gaussian elimination method for solving these equations is usually taught in an elementary algebra course. The method proceeds as follows:

Divide the first equation by $a_{1,1}$ to make the first coefficient equal to 1. Multiply this new first equation by $a_{2,1}$, and subtract the result from the second equation. Multiply the new first equation by $a_{3,1}$, and subtract this result from the third equation, and so forth. In $n - 1$ steps we get $n - 1$ equations, none of which have the first variable x_1, but have only the variables $x_2, x_3, \ldots, x_n$.

Solve

(1)	$3x + 6y + 9z = 39$
(2)	$2x + 5y - 2z = 3$
(3)	$x + 3y - z = 2$

Divide (1) *by* 3

(1′)	$x + 2y + 3z = 13$
(2′)	$2x + 5y - 2z = 3$
(3′)	$x + 3y - z = 2$

Subtract twice (1') from (2') and (1') from (3')

(1")	$x + 2y + 3z = \quad 13$
(2")	$y - 8z = -23$
(3")	$y - 4z = -11$

Divide (2") by 1, and subtract (2") from (3")

(1''')	$x + 2y + 3z = \quad 13$
(2''')	$y - 8z = -23$
(3''')	$4z = \quad 12$

Next we repeat this process and eliminate the variable x_2 in $n - 2$ steps. And so forth.

If all goes well, we finally come down to a single equation in x_n which is easily solved.

We then substitute this value of x_n in the equation having only x_n and x_{n-1} and solve for x_{n-1}.

Back substitution

$$z = 3$$
$$y = -23 + 24 = 1$$
$$x = 13 - 2 - 9 = 2$$

Hence,

$$x = 2$$
$$y = 1$$
$$z = 3$$

is the solution.

The obvious repetition of this **back-substitution** process produces the solution, one x_i at a time in the sequence $x_n, x_{n-1}, \ldots, x_1$. Thus the system of equations is solved.

In this process we have divided an equation by its leading coefficient n times,

The divisors were

$$3, 1, 4$$

Hence the determinant has the value

$$3 \times 1 \times 4 = 12$$

and each of these divisions changes the value of the determinant by the corresponding amount. The final determinant clearly has the value of 1, and since (by Theorem 5 of Appendix A) none of the other operations changed the value of the determinant of the system,

---

Product notation

$$\prod_{i=1}^{n} x_i = x_1 x_2 \cdots x_n$$

Determinant

$$|a_{ij}| = \prod_{i=1}^{n} d_i$$

where the d_i are the divisors used.

---

it follows that the value of the original determinant is the product of the divisors we used.

PROBLEMS 5.1

1 Solve the system

$$x + y + z = 6$$
$$2x + 3y + z = 1$$
$$x - y + z = 3$$

What is the value of the determinant?

2 Solve the system

$$x - y + z = 1$$
$$2x + 2y - z = 3$$
$$3x - 3y + z = 1$$

3 Solve the system

$$ax + by = e$$
$$cx + dy = f$$

4 Count the number of additions, multiplications, and divisions in the total Gaussian elimination process as described above.

5.2 Pivoting

If the solution of simultaneous linear equations is this simple, then why is a large part of a chapter devoted to this topic? The answer is that we have neglected a number of important details, including the difficult topic of roundoff errors.

Gaussian elimination with pivoting, either partial or complete, is the generally recommended method for solving linear equations. By **partial pivoting** we mean the selection of the largest-sized coefficient in the next column and the use of the corresponding equation as a basis for the elimination process.

(1)	$x + y + z = 6$
(2)	$2x - y + z = 3$
(3)	$3x + 2y - z = 4$

$\llcorner$*First pivot*

$(3')$ $\qquad x + \frac{2}{3}y - \frac{1}{3}z = \frac{4}{3}$

Subtract $(3')$ from (1) and twice $(3')$ from (2)

$(1')$ $\qquad \frac{1}{3}y + \frac{4}{3}z = \frac{14}{3}$

$(2')$ $\qquad -\frac{7}{3}y + \frac{5}{3}z = \frac{1}{3}$

$\llcorner$*Second pivot*

$(2'')$ $\qquad y - \frac{5}{7}z = -\frac{1}{7}$

Multiply $(2'')$ by $\frac{1}{3}$, and subtract from $(1')$

$(1''')$ $\qquad \frac{33}{21}z = \frac{99}{21}$

$\qquad z = 3$

Put this in $(2'')$

$$y = -\frac{1}{7} + \frac{15}{7} = \frac{14}{7} = 2$$

Put both in (1)

$$x = 6 - 2 - 3 = 1$$

Solution

$$\begin{cases} x = 1 \\ y = 2 \\ z = 3 \end{cases}$$

This avoids the difficulty of what to do if the coefficient we wish to divide by happens to be zero (but see later if all are zero). **Complete pivoting** is the selection of the largest-sized of all the available coefficients as the basis for the next stage of elimination, which, of course, operates on the corresponding variable. It is generally agreed that it is usually not worth the extra search time to do complete pivoting.

The effect of doing partial pivoting can be regarded as merely a reordering of the sequence of the equations in the given system, though this should not actually be done in the machine because, among other things, it would greatly confuse the user if he tries to find out what happened inside the machine.

5.3 Rank

The concept of congruent triangles in high-school geometry does not make a distinction between left-handed and right-handed triangles, though this is important in some applications of geometry. Similarly, the usual definition of the rank of a determinant (Appendix A, Sec. 5A.4) does not distinguish between row and column dependence (in a certain sense they are equivalent, but there are practical differences).

Column dependence

$$2x + 4y + z = 13$$
$$Pivot\nearrow \quad x + 2y - z = 2$$
$$x + 2y + 2 = 11$$

$$x + 2y + \tfrac{1}{2}z = \tfrac{13}{2}$$
$$0y - \tfrac{3}{2}z = -\tfrac{9}{2}$$
$$0y + \tfrac{3}{2}z = \tfrac{9}{2}$$

Column of all zeros

Row dependence

$$2x + 4y + z = 13$$
$$Pivot\nearrow \quad x + 2y + \tfrac{1}{2}z = \tfrac{13}{2}$$
$$x + y + z = 6$$

$$x + 2y + \tfrac{1}{2}z = \tfrac{13}{2}$$
$$0y + 0z = 0$$
$$-y + \tfrac{1}{2}z = -\tfrac{1}{2}$$

Row of all zeros

This distinction is often important to the user of a library routine when he wishes to understand what has happened to his problem if the system does not have the maximum rank.

How are we to recognize these two kinds of linear dependence? In the partial pivoting process, if we find in searching for a pivot in the next column of numbers that all of the numbers are zero, then this clearly (with a little thought perhaps) shows that the column of zeros is some linear combination of the preceding columns. If instead we were to find a row of zeros, then this indicates that we have a linear dependence among the rows that we have so far used. Thus, in the process of searching for our next pivot, we need to check each zero in the column to see if it is part of an entire row of zeros, as well as to check to see if the entire column is zeros.

Having found a linear dependence, we should try to locate it as closely as we can so that the user is pointed in the right direction, since it is highly probable that he intended the equations to have the maximal rank. To locate the dependence more closely, we can drop one row† (column) at a time and restart to see if the same row (column) of zeros appears. If it does not, then the dropped equation (variable) is part of the linear dependence; if it does, then we drop the row (column) and try dropping another. In time we are led to a minimal set of equations having the linear dependence.

In principle there can be arbitrarily difficult combinations of row and column dependence and an arbitrarily large decrease in rank, but in practice this is unusual because, as we have said, the user generally intends the system to be of the maximal rank so we shall not go further into this intricate topic.

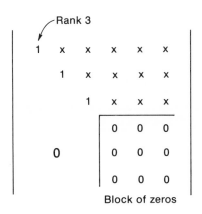

The relevant mathematical theorem (which we shall not bother to prove) is:

If the rank r of the matrix of coefficients of the unknowns is the same as the rank of the augmented matrix, then r of the equations can be solved for in terms of the other n − r variables; and if the ranks are not the same, then the equations are inconsistent.

---

†In this notation, read always the word before the parentheses or always the word in the parentheses.

It is perhaps worth noting, especially for column linear dependence, that many of the variables can be solved for uniquely and only some of the variables have a degree of arbitrariness.

PROBLEMS 5.3

1 Solve

$$x + y + z = 6$$
$$3x - y + z = -2$$
$$2x + 0y + z = 2$$

2 Solve

$$x + 3y + 2z = 10$$
$$x - y + 0z = -1$$
$$x + y + z = 6$$

3 Solve

$$x + y + z = a$$
$$x + y + z = a$$
$$x + y + z = a$$

5.4 Roundoff

The main difficulty with the preceding discussion is that it ignored the vital problem of roundoff.

Given the system

$$x + \tfrac{1}{3}y = 1$$

$$2x + \tfrac{2}{3}y = 2$$

The rank is 1.

Put the system in a computer:

$$\text{Note coefficients}$$

$$.100 \times 10^1 x + .333 \times 10^0\, y = .100 \times 10^1$$

$$.200 \times 10^1 x + .667 \times 10^0 y = .200 \times 10^1$$

Now the rank is 2, and we can solve the system uniquely!

$$.100 \times 10^{-2} y = 0$$
$$\begin{cases} y = 0 \\ x = 1 \end{cases}$$

As shown by the trivially simple example of rank 1 in the insert, as soon as we commit roundoff, we raise the rank of the system to the full amount. Add to the roundoff of the initial coefficients, the conversion to binary and the subsequent arithmetic, and it becomes clear that we cannot in general expect to find the zeros we discussed in the section on rank.

We therefore face the question of how small is "small." Unfortunately, the answer is not easy to give. For one reason, once the equations are fixed, then the value of the determinant is also fixed and the product of all the divisors (which are the **largest** numbers in their respective columns) is also determined. Therefore, the more successful we are in finding large pivots in the beginning, the more sure we are that we must find compensating small ones later on. In short, in the process of Gaussian elimination there are large-scale internal correlations in the arithmetic, and we are unable to follow them through in a simple fashion so that we can give a simple meaning to the idea of "small."

When we suspect a rank less than n, we can artificially try various (multiplicative) levels of roundoff noise and call numbers less than this noise zero, but there appears to be no way of having the routine in the machine make reasonable judgments on this matter and then use them automatically.

5.5 Scaling

The purpose of preliminary scaling is, among other things, to give meaning to "large" and "small."

As for polynomials, we need to examine the question of equivalent systems that can arise from scaling; unless we can somehow scale the equations properly, then what significance can there be in the statement, "take the largest number in the column as the pivot," other than the chance of the way the equations were written down?

Consider the matrix of coefficients

$$\begin{pmatrix} 3 & \dfrac{1}{\varepsilon} & \dfrac{1}{\varepsilon} \\ 1 & 2\varepsilon & \varepsilon \\ 2 & \varepsilon & \varepsilon \end{pmatrix}$$

where we assume that ε is small and that

$$1 \pm \varepsilon^2 = 1$$

Scale by rows

$$\begin{pmatrix} 3\varepsilon & 1 & 1 \\ 1 & 2\varepsilon & \varepsilon \\ 2 & \varepsilon & \varepsilon \end{pmatrix}$$

Pivot

Pivot

$$2\begin{pmatrix} 0 & 1 & 1 \\ 0 & 3\dfrac{\varepsilon}{2} & \dfrac{\varepsilon}{2} \\ 1 & \dfrac{\varepsilon}{2} & \dfrac{\varepsilon}{2} \end{pmatrix}$$

$$2\begin{pmatrix} 0 & 1 & 1 \\ 0 & 0 & -\varepsilon \\ 1 & \dfrac{\varepsilon}{2} & \dfrac{\varepsilon}{2} \end{pmatrix}$$

and the back substitution would be easy.

But *if we scale first by columns,*

Pivot

$$\begin{pmatrix} 3 & 1 & 1 \\ 1 & 2\varepsilon^2 & \varepsilon^2 \\ 2 & \varepsilon^2 & \varepsilon^2 \end{pmatrix}$$

$$\begin{pmatrix} 3 & 1 & 1 \\ 0 & -\dfrac{2}{3} & -\dfrac{1}{3} \\ 0 & -\dfrac{2}{3} & -\dfrac{2}{3} \end{pmatrix}$$

The rank is 2.

It is obvious that there are n scale factors that can be used on the rows and another n on the variables by using

$$\bar{x}_i = k_i x_i$$

without changing the system in any fundamental way. We could, if we wish, also scale the constants on the right-hand side, but this is seldom worth the trouble. We therefore have $2n$ scale factors to decide upon.

About all the help in scaling one can find in the standard textbooks is something like "scale by rows and then by columns (or the other way around) to get the largest element in any row and any column to be around 1 in size." The simple example above shows the superficiality of this advice. Evidently, the usual advice about scaling is inadequate, and we are in trouble if we try to defend the pivoting method in detail.

PROBLEMS   5.5

Scale as best you can:

1 $x + 10y + 100z = 3$
       $10x - y + 10z = 72$
       $100x + 100y + z = 9$

2 $100x + 10y + z = 100$
       $100x - y + z = 10$
       $100x + y - z = 1$

3 $100x + y + z = 72$
       $x + 100y + z = 9$
       $x + y + 100z = -7$

5.6   A method of simultaneous row and column scaling

Since we use multipliers of the rows and columns as scaling factors, it is natural to look at the logarithms of the absolute values of the coefficients. Let us imagine that the rows have been multiplied by 2^{r_i} $(i = 1, \ldots, n)$ and the columns by 2^{c_j} $(j = 1, \ldots, n + 1)$, where r_i and c_j need not be integers.† We shall also multiply all of the $n(n + 1)$ elements of the system by 2^M. We write the augmented matrix

$$\begin{pmatrix} a_{1,1} & \cdots & a_{1,n} & b_1 \\ \cdots\cdots\cdots\cdots\cdots\cdots \\ a_{n,1} & \cdots & a_{n,n} & b_n \end{pmatrix} \equiv A$$

and set

$$|a_{i,j}| = 2^{b_{ij}} \qquad \begin{array}{l} (i = 1, \ldots, n) \\ (j = 1, \ldots, n + 1) \end{array}$$

---

†It may well be preferable to scale only the first n of the $n + 1$ columns since the constant terms are not involved in the selection of pivots. If so, the modifications are easily made.

where $a_{i,n+1} \equiv b_i$. Supposing for the moment that $a_{i,j} \neq 0$, then the new exponent is

$$b_{i,j} + M + r_i + c_j$$

In order to make all $n(n + 1)$ new elements simultaneously as close to 0 as possible, we will **arbitrarily** minimize

$$m = \sum_{i=1}^{n} \sum_{j=1}^{n+1} (b_{i,j} + M + r_i + c_j)^2$$

We have more than enough parameters, and we select

$$M = -\frac{1}{n(n + 1)} \sum_{i=1}^{n} \sum_{j=1}^{n+1} b_{i,j}$$

as the negative of the average of all the $b_{i,j}$. As the first step in finding the minimum, we differentiate with respect to the variables r_i and c_j, and set them equal to zero to get

$$\frac{\partial m}{\partial r_i} = 2 \sum_{j=1}^{n+1} (b_{i,j} + M + r_i + c_j) = 0 \qquad (i = 1, \ldots, n)$$

$$\frac{\partial m}{\partial c_j} = 2 \sum_{i=1}^{n} (b_{i,j} + M + r_i + c_j) = 0 \qquad (j = 1, \ldots, n + 1)$$

These equations have the solution

$$r_i = \frac{-1}{n + 1} \sum_{j=1}^{n+1} (b_{i,j} + M)$$

hence,

$$\sum_{i=1}^{n} r_i = 0$$

$$c_j = \frac{-1}{n} \sum_{i=1}^{n} (b_{i,j} + M)$$

hence,

$$\sum_{j=1}^{n+1} c_j = 0$$

Direct substitutions show that these are the solutions since

$$2[-(n+1)r_i + (n+1)r_i + 0] = 0$$

$$2(-nc_j + 0 + nc_j) = 0$$

Thus, the r_i and c_j are the appropriate averages of

$$b_{i,j} + M$$

This is the classical analysis of variance.

If any of the coefficients $a_{i,j} = 0$, then there is no corresponding $b_{i,j}$ and we must treat it as "missing data" as discussed in conventional statistics books on the topic.

In this approach to scaling, we have minimized the variance of the scaled exponents of the terms that are in the system (and have excluded the zero terms). As usual when we come to carry out the scaling, we calculate the scale factors to use in picking out the pivots, but we do **not** scale the equations themselves, so that the question of integer solutions to the scaling equations need not arise. In practice it may be sufficient merely to use the exponents of the $a_{i,j}$ as the $b_{i,j}$ without taking the full logarithm.

Unfortunately, this method of scaling cannot be shown to be relevant to any known method of solution; it is merely a plausible method to use in place of the usual "scale by rows and then by columns" rule.

PROBLEMS 5.6

1 Scale Prob. 3 in Sec. 5.5.
2 Scale Prob. 1 in Sec. 5.3.
3 Scale Prob. 1 in Sec. 5.5.

5.7 Ill-conditioned systems

Since we have failed to be precise both in identifying zeros so that we can determine the rank of the system and in finding a sound meaning to scaling, it is necessary to find some concept that will at least do part of the job. This idea is the "ill-conditioned system." The term "ill-conditioned" is ill defined. The vague idea is that small changes in the initial system can produce large changes in the final result. If we are to take floating point seriously, then we should say "relatively small changes" and "relatively large changes." Thus, in a sense the idea of ill conditioning is a substitute for the idea of rank in the presence of roundoff. Alternatively, it can be said to be an attempt to cope with the idea of linear dependence in the presence of roundoff noise. We seem to need some idea like "almost linearly dependent," but as yet this idea has not been formulated clearly.

If the ill-conditioned effect is in the original physical system, then it is usually called "unstable." Thus, a pencil balanced on its point is an unstable system since small changes in the initial position result very soon in large differences in the subsequent positions of the pencil.

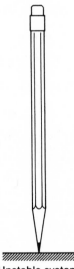

Unstable system

The ill conditioning may arise not from the original physical system but from the mathematical formulation of the problem. In a sense the material on function evaluation in Chap. 1 shows how different formulations of mathematically equivalent equations can lead to accurately computable or to very inaccurately computable expressions. Another example is in the choice of the basis for representing the function (see Chap. 12 for still another example of this). Thus, if for the differential equation

$$\frac{d^2y}{dx^2} = y \quad \begin{cases} y(0) = 1 \\ y'(0) = -1 \end{cases}$$

we choose the hyperbolic functions as our basic solutions, we get the solution

$$y = \cosh x - \sinh x$$

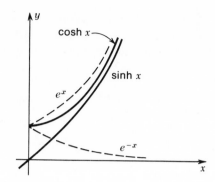

whereas if we choose the exponential functions (which are mathematically completely equivalent), we get

$$y = e^{-x}$$

For large values of x the second is easily calculated, whereas the first is impractical. Yet another example is in the classic case of the circular vibrating drumhead. If we introduce cylindrical coordinates, as is customary, we find that the differential equation we have to solve has a singularity at the origin—a singularity that is not present in the original problem!

Bessel's equation

$$\frac{d^2y}{dx^2} + \frac{1}{x}\frac{dy}{dx} + y = 0$$

Singularity is due to the coordinate system and not to the physical problem.

Thus ill conditioning can be a result of the matematical formulation of the problem, and if so, this can best be handled by a reexamination of the mathematical steps rather than by any attempts to get through somehow by double or triple precision or other fancy gimmicks.

Even when the problem is formulated in a reasonable way, the process of solution may make it unstable. Using only the vague definition above, we announce the simple theorem:

Theorem   *Pivoting can take a well-conditioned system into an ill-conditioned system of simultaneous linear equations.*

Proof   The example of scaling in Sec. 5.5 shows that the ill conditioning arose not in the scaling but in the choice of the top line to
eliminate from the other two lines.

The symmetric system

$$\begin{array}{r} \rightarrow\; 3x + 2y + z = 3 + 3\varepsilon \\ Pivot\;\; 2x + 2\varepsilon y + 2\varepsilon z = 6\varepsilon \\ x + 2\varepsilon y - \varepsilon z = 2\varepsilon \end{array}$$

appears to be properly sealed, so we pivot to get

$$(-\tfrac{4}{3} + 2\varepsilon)y + (-\tfrac{2}{3} + 2\varepsilon)z = -2 + 4\varepsilon$$
$$(-\tfrac{2}{3} + 2\varepsilon)y + (-\tfrac{1}{3} - \varepsilon)z = -1 + \varepsilon$$

*For small ε, these are ill conditioned. **But** if we eliminate between the second
and third equations, we get*

$$\varepsilon y - 2\varepsilon z = -\varepsilon$$

In floating point, we divide out the ε,

$$y - 2z = -1$$

and can now solve for

$$y = z = 1$$

*Put in the first equation, we have trouble, but it is safe in either of the others, to
get*

$$x = \varepsilon$$

The example in the insert proves the theorem for simultaneous equations with the use of a simple symmetrical, 3×3 system.

    There are a number of misconceptions about ill-conditioned sys-
tems. One misconception is that once the system is scaled, then a
small determinant of the system must mean it is ill conditioned, but as
we have just seen, this is not so.

    Another misconception is that somehow ill conditioned is con-
nected with "the angle between the graphs of the equations." However, in floating point arithmetic, a scale change of $x = \lambda x$ can change
the angle radically **without** changing significantly the arithmetic we
shall do to get the solution.

PROBLEM 5.7

1 Given the lines

$$y = m_1 x + b_1$$
$$y = m_2 x + b_2$$

show that if we scale $x = \lambda \bar{x}$, then there is a λ such that for $m_1 m_2 < 0$, the lines are perpendicular. For $m_1 m_2 > 0$, find the λ that makes the angle greatest.

5.8   Why did the examples work?

It is reasonable to ask how typical these examples are and how often in the past the pivoting method has created the ill conditioning that was reported to occur by some library routines. The answers are not known at this time; all that is claimed is that textbooks and library descriptions rarely, if ever, mention this possibility (though it is apparently known in the folklore).

When we did the pivoting in the examples, we added to the two lower equations the large numbers in the y and z columns of the top equation.

If in the earlier example we set $x = \varepsilon \bar{x}$, the equations become

$$3\varepsilon \bar{x} + 2y + \quad z = 3 + 3\varepsilon$$
$$2\bar{x} + 2y + 2z = 6$$
$$\bar{x} + 2y - \quad z = 2$$

and we would never choose the 3ε coefficient as the pivot. Thus, in a sense the example depended on the "wrong units" in x.

Thus, we later found a "linear dependence" because of the finite size of our computing system. Indeed, it is now easy to see that in general the use of a pivot for the elimination of a variable from all the other equations leads us to a set of equations where we "see" the original pivoting equation everywhere we look in the derived system; we are apparently trying hard to make the system linearly dependent!

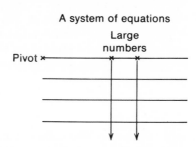

A system of equations

But what of the well-known arguments that "prove" that pivoting is a good thing because it tends to reduce roundoff?

Pivoting produces effective multipliers of the pivoting equation that are less than 1 in size; hence it apparently does not amplify the earlier roundoffs.

If L_i means a typical equation and $|\lambda| < 1$,

$$\lambda L_1 + L_2 = L_3$$

is a typical step in the elimination process. But

$$L_1 + \frac{1}{\lambda} L_2 = \frac{1}{\lambda} L_3$$

has the same relative roundoff!

When we examine this, we find (using the jargon of chess-playing programs) that with "zero look ahead" we are minimizing the current roundoff propagation. If we care to look merely one step ahead, we can see that we may be inviting linear dependence. And what is so easily seen in 3×3 systems can only be imagined in all its subtlety in higher-order systems. It would appear that more than a "zero look ahead" is necessary if we are to avoid the error of the examples.

5.9 Ill conditioning can depend on the right-hand side

Much of the material in textbooks on ill-conditioned systems refers to properties of the matrix of coefficients of the unknowns. But if we change the right-hand side in the example of simultaneous equations to get the system

$$\begin{aligned}
3x + 2y + z &= 6 \\
2x + 2\varepsilon y + 2\varepsilon z &= 2 + 4\varepsilon \\
x + 2\varepsilon y - \varepsilon z &= 1 + \varepsilon
\end{aligned}$$

then the solution will be $x = y = z = 1$ and the system is indeed ill conditioned because, no matter how we try, we are unable to solve the system so that the answer is not sensitive to small changes in the original coefficients. In a sense we can find no way to get to the key step of dividing out the ε that does not require the exact cancellation of large factors before we get to the division.

5.10 Gaussian elimination again

We have spent a good deal of time exposing the weaknesses of the Gaussian elimination process because so many library routines are based on it and hence you may be led astray if you believe everything that you get from the machine. We have not justified the pivoting method; rather we have shown that it is an "old wives' tale." But like most old wives' tales, it is a mixture of truth and mystic faith.

Experimental evidence shows that there is some truth in the rule that you should pick the biggest pivot you can in the next column. How can this be? It is probably nothing more than the simple observation that small numbers can arise **either** by multiplying by small numbers **or** by large cancellations which leave little accuracy, whereas large numbers can arise **only** from multiplying by large factors and not by additions. Thus, other things being the same, large pivots are probably safer than small ones which may occasionally have only a little accuracy in them and hence should be avoided.

5.11 The Gauss-Jordan process

Instead of doing the back substitution, we may in principle reduce the left-hand side to a strictly diagonal form by eliminating all the off-diagonal terms. This may be done as follows: First do the elimination process as before to obtain all zeros below the main diagonal. Then use the last equation to eliminate all the x_n in the remaining $n - 1$ equations. Then use the next-to-the-last equation to eliminate all the x_{n-1} in the remaining $n - 2$ equations. And so forth.

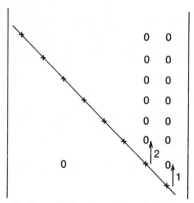

Further elimination of coefficients
of the unknowns

This method involves too much arithmetic to justify ever doing it, even if there are many right-hand sides all with the same left-hand side!

5.12 The inverse matrix—1

In arithmetic the problem of division

$$\frac{b}{a}$$

can be reduced to that of finding the "inverse," or reciprocal,

$$a^{-1} = \frac{1}{a}$$

and then computing the product

$$\left(\frac{1}{a}\right)b = a^{-1}b$$

With matrices, the same may be done **except** that we must be careful because AB is not necessarily BA and we must consider both a left and a right inverse

$$A_L^{-1} A = 1$$
$$AA_R^{-1} = 1$$

It is an important mathematical fact that

$$A_R^{-1} = A_L^{-1}$$

This can be seen by considering (assuming that they exist)

$$A_L^{-1}AA_R^{-1} = \begin{cases} A_L^{-1}(AA_R^{-1}) = A_L^{-1}I = A_L^{-1} \\ (A_L^{-1}A)A_R^{-1} = IA_R^{-1} = A_R^{-1} \end{cases}$$

But in the presence of roundoff this may be far from true!

---

because of the associativity of matrix multiplication.

One approach to the inverse of a matrix is based on the idea of solving many systems of linear equations, all of which have the same left-hand side but differ in the right-hand side. This idea means that if we could find the solutions for the special right-hand sides

$$e_1 = \begin{pmatrix} 1 \\ 0 \\ 0 \\ \vdots \\ 0 \end{pmatrix}, \quad e_2 = \begin{pmatrix} 0 \\ 1 \\ 0 \\ \vdots \\ 0 \end{pmatrix}, \quad e_3 = \begin{pmatrix} 0 \\ 0 \\ 1 \\ \vdots \\ 0 \end{pmatrix}, \quad \cdots, \quad e_n = \begin{pmatrix} 0 \\ 0 \\ 0 \\ \vdots \\ 1 \end{pmatrix}$$

then we could multiply these solutions by the given right-hand side coefficients b_i and add them to get the solution. These multiplications followed by additions can be written as some matrix times the column matrix b. The columns of the first matrix are exactly the solutions to the special right-hand sides. We therefore have the solution to the given system in the form

$$x = (\ \ )b$$

We started with the matrix equation

$$Ax = b$$

and if we had multiplied both sides of the equation on the left by the inverse matrix A^{-1}, then we should have exactly what we found above. Thus, the matrix of the special solutions is the inverse matrix.

5.13 The inverse matrix—2

Because the inverse matrix plays such a prominant role in computing, we shall give a second approach to the matter, one that leads to exactly the same way of computing the inverse.

In the second method we start with the matrix of size $n \times 2n$

$$(A, I) = \begin{pmatrix} a_{1,1} & a_{1,2} & \cdots & 1 & 0 & \cdots \\ a_{2,1} & a_{2,2} & \cdots & 0 & 1 & \cdots \\ \cdots\cdots\cdots\cdots\cdots\cdots\cdots\cdots\cdots\cdots\cdots \end{pmatrix}$$

Let us apply the usual Gaussian elimination to this matrix, operating on the complete rows. After reducing the leftmost part to an upper diagonal matrix whose lower left-hand corner is all zeros, we continue with the Gauss-Jordan back substitution (Sec. 5.11) and reduce the leftmost n columns to the unit matrix. In this process we have multiplied rows by constants, which is the same as multiplying on the left by the matrix

$$\begin{pmatrix} 1 & 0 & \cdots & 0 & \cdots & 0 \\ 0 & 1 & \cdots & 0 & \cdots & 0 \\ \cdots\cdots\cdots\cdots\cdots\cdots\cdots\cdots\cdots \\ 0 & 0 & \cdots & k & \cdots & 0 \\ \cdots\cdots\cdots\cdots\cdots\cdots\cdots\cdots\cdots \\ 0 & 0 & \cdots & 0 & \cdots & 1 \end{pmatrix}$$

We have also multiplied a row by a constant k and added this to another row, which is the matrix of the form

$$\begin{pmatrix} 1 & 0 & \cdots & 0 & \cdots & 0 \\ 0 & 1 & \cdots & 0 & \cdots & 0 \\ & & \cdots\cdots\cdots\cdots\cdots & & & \\ k & 0 & \cdots & 1 & \cdots & 0 \\ & & \cdots\cdots\cdots\cdots\cdots & & & \\ 0 & 0 & \cdots & 0 & \cdots & 1 \end{pmatrix}$$

All these matrices are applied to the left-hand side A and **also** at the same time to the right-hand side I. If we gather all these matrices together and multiply them to form a single matrix (which we may do by the associativity of matrix multiplication), then we have the matrix which converts A to I and thus we have A^{-1}. But in the process this is also multiplied in the right-hand n columns of the matrix, namely, I, and we therefore find in the right-hand n columns of the product of A^{-1} times I, which is exactly A^{-1}.

A little thought will reveal that this is exactly what we were doing in the previous section when we found the simple solutions to the special right-hand sides. Thus, as far as computing is concerned, both methods are the same; they are merely alternative ways of looking at the process.†

PROBLEMS   5.12

Find the inverse of:

1 $\begin{pmatrix} 1 & 2 & 3 \\ 4 & 5 & 6 \\ 7 & 8 & 9 \end{pmatrix}$

2 $\begin{pmatrix} 1 & -1 & 1 \\ 1 & 0 & 1 \\ 1 & 2 & 1 \end{pmatrix}$

3 $\begin{pmatrix} 1 & 2 & 1 \\ 2 & 1 & 2 \\ 1 & 1 & 1 \end{pmatrix}$

---

†For further information on the topic of this chapter, see Ben Noble, "Applied Linear Algebra," Prentice-Hall, Inc., Englewood Cliffs, N.J., 1969.

APPENDIX: A BRIEF INTRODUCTION TO DETERMINANTS

5A.1 The definition of a determinant

Definition *The value of a determinant*

$$D = |a_{i,j}|$$

is the sum of all possible products which can be formed by taking one element from each row and each column and attaching the proper sign.

The proper sign can be found by writing the individual $a_{i,j}$ terms in each product so that the first subscripts are in the order 1, 2, 3, ..., n,

$$a_{1,j_1} a_{2,j_2} a_{3,j_3} \cdots a_{n,j_n}$$

and then counting the number of interchanges necessary to put the second subscripts into the order 1, 2, 3, ..., n. If the number of interchanges is even, then the sign is $+$; if odd, then the sign is $-$.

Example

Consider the term

$$a_{1,3} a_{2,2} a_{3,1}$$

One interchange

does the job of ordering the second subscript; thus, the term has a minus sign.

Another way is

$$
\begin{array}{ccc}
3 & 2 & 1 \\
3 & 1 & 2 \\
1 & 3 & 2 \\
1 & 2 & 3
\end{array}
$$

Again, the sign is minus.

The trouble with this definition, besides being long and compli-cated, is that it is not obvious that it is a definition. Is it clear that no matter how we do the interchanges we will always get the same sign for the same term?

Another example

Hence +

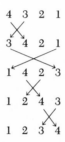

Again +

To examine this point, consider the function $P = P(x_1, x_2, \ldots, x_n)$ defined by

$$P = (x_1 - x_2)(x_1 - x_3) \cdots (x_1 - x_n)(x_2 - x_3) \cdots (x_2 - x_n) \cdots (x_{n-1} - x_n)$$

$$= \prod_{i<j=2}^{n} (x_i - x_j)$$

where all the x_k are distinct. Now P has the following property: If we interchange an x_i with an x_j, then we change the sign of P. To see this, suppose we interchange x_i and x_j. For each term $x_k - x_i$, for $k \neq i$ or j, there is a term $x_k - x_j$, and the product of these two terms $(x_k - x_i)(x_k - x_j)$ does not change sign. But there is also a term $x_i - x_j$ which becomes $x_j - x_i$, and hence P has a change in sign when we in-terchange x_i and x_j.

If we start with the sequence of second subscripts

$$j_1, j_2, j_3, \ldots, j_n$$

and by m interchanges bring the sequence to 1, 2, 3, . . . , n, then we shall have a total change of sign in P of $(-1)^m$. If we also get to the final sequence by another route using t interchanges, then we shall have a sign change of $(-1)^t$. But the beginning and ending products are the same, so that

$$(-1)^m = (-1)^t$$

Therefore $t = m + 2k$ for some k, and we have proved that the number of interchanges to go from any sequence of subscripts to the standard sequence is either even or odd **independent** of the particular path taken. Thus, our definition of a determinant is unique.

5A.2 Some simple properties

A little thought shows that if we had started with the second subscript in the correct order and made the interchanges until the first subscript was in order, then we should naturally have the same sign on the term (we are reversing the path of the interchanges). This shows that the roles of columns and rows can be interchanged without changing the value of the determinant. A matrix (rectangular array of numbers, see Appendix 5B) with rows and columns interchanged is called the **transposed matrix**.

Original

$$\begin{vmatrix} 1 & 2 & 3 \\ 4 & 5 & 6 \\ 7 & 8 & 9 \end{vmatrix}$$

Transpose

$$\begin{vmatrix} 1 & 4 & 7 \\ 2 & 5 & 8 \\ 3 & 6 & 9 \end{vmatrix}$$

Thus, we have the **metatheorem** that any theorem we prove for the rows of a determinant is automatically true for the columns, and we shall not bother to state the second theorem each time.

Theorem 1   *If we multiply the elements in any row by some constant k, then we have multiplied the value of the determinant by k.*

The proof follows immediately from the definition since the factor k will appear exactly once in each term of the sum of products which is the value of the determinant.

$$\begin{vmatrix} 1 & 2 & 3 \\ 4k & 5k & 6k \\ 7 & 8 & 9 \end{vmatrix} = k \begin{vmatrix} 1 & 2 & 3 \\ 4 & 5 & 6 \\ 7 & 8 & 9 \end{vmatrix}$$

Theorem 2   *The interchange of any two rows of a determinant changes the sign of the value of the determinant.*

The proof follows from the definition since the interchange of two rows produces exactly one interchange of subscripts in each term of the sum, hence a change of sign in each term.

$$\begin{vmatrix} 1 & 2 & 3 \\ 4 & 5 & 6 \\ 7 & 8 & 9 \end{vmatrix} \quad Interchange \quad = - \begin{vmatrix} 7 & 8 & 9 \\ 4 & 5 & 6 \\ 1 & 2 & 3 \end{vmatrix}$$

Theorem 3   *If two rows of a determinant are the same, then the value D of the determinant is zero.*

From Theorem 2, the interchange of the identical rows changes the sign so that

$$D = -D$$

which means that $D = 0$.

$$\begin{vmatrix} a & b & c \\ a & b & c \\ x & y & z \end{vmatrix} = 0$$

for any x, y, z, a, b, c.

Corollary   *If two rows of a determinant are proportional to each other, then the value of the determinant is zero.*

By Theorem 1 we can multiply one of the two rows by the proper constant to make the two rows equal and then apply Theorem 3.

$$\begin{vmatrix} a & b & c \\ ka & kb & kc \\ x & y & z \end{vmatrix} = 0$$

Theorem 4 *If each element of the ith row is written as a sum*

$$a_{i,j} = a'_{i,j} + a''_{i,j}$$

then the determinant is the sum of two determinants

$$\begin{vmatrix} a_{1,1} & a_{1,2} & \cdots & a_{1,n} \\ \cdots\cdots\cdots\cdots\cdots \\ a_{i,1} & a_{i,2} & \cdots & a_{i,n} \\ \cdots\cdots\cdots\cdots\cdots \\ a_{n,1} & a_{n,2} & \cdots & a_{n,n} \end{vmatrix} = \begin{vmatrix} a_{1,1} & a_{1,2} & \cdots & a_{1,n} \\ \cdots\cdots\cdots\cdots\cdots \\ a'_{i,1} & a'_{i,2} & \cdots & a'_{i,n} \\ \cdots\cdots\cdots\cdots\cdots \\ a_{n,1} & a_{n,2} & \cdots & a_{n,n} \end{vmatrix}$$

$$+ \begin{vmatrix} a_{1,1} & a_{1,2} & \cdots & a_{1,n} \\ \cdots\cdots\cdots\cdots\cdots\cdots \\ a''_{i,1} & a''_{i,2} & \cdots & a''_{i,n} \\ \cdots\cdots\cdots\cdots\cdots\cdots \\ a_{n,1} & a_{n,2} & \cdots & a_{n,n} \end{vmatrix}$$

Again the proof follows immediately from the definition since each product from the determinant on the left-hand side corresponds to a pair of products, one from each of the right-hand side determinants.

Typical terms

$$a_{1,j_1} \cdots (a'_{i,j_i} + a''_{i,j_i}) \cdots a_{n,j_n}$$

$$= a_{i,j} \cdots a'_{i,j_i} \cdots a_{n,j_n} + a_{1,j_i} \cdots a''_{i,j_i} \cdots a_{n,j_n}$$

Theorem 5 *If any row of a determinant is multiplied by a constant k and is added term by term to another row of the determinant, then the value of the determinant is not changed.*

To prove this apply Theorem 4 to the resulting determinant to get two determinants, the original determinant and one with a pair of rows which are proportional. The latter one has a value of zero by the corollary of Theorem 3.

$$\begin{vmatrix} 1 & 2 & 3 \\ 4 & 5 & 6 \\ 7 & 8 & 9 \end{vmatrix}$$

Multiply the top row by 4 and subtract from the second row

$$= \begin{vmatrix} 1 & 2 & 3 \\ 0 & -3 & -6 \\ 7 & 8 & 9 \end{vmatrix}$$

This theorem shows some of the relevance of determinant theory to Gaussian elimination.

5A.3 Minors

Definition   *If in a given $n \times n$ determinant*

$$|a_{i,j}|$$

we delete the ith row and jth column and regard what is left as an $(n-1) \times (n-1)$ determinant, we call this the **minor** *of the element $a_{i,j}$ and label the minor*

$$A_{i,j}$$

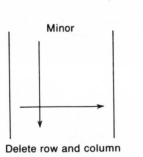

Minor

Delete row and column

Theorem 6   *The determinant D may be written as*

$$a_{11}A_{11} - a_{12}A_{12} + a_{13}A_{13} - \cdots + (-1)^{n-1}a_{1n}A_{1n}$$

By the first row

$$\begin{vmatrix} 1 & 2 & 3 \\ 4 & 5 & 6 \\ 7 & 8 & 9 \end{vmatrix} = 1 \begin{vmatrix} 5 & 6 \\ 8 & 9 \end{vmatrix} - 2 \begin{vmatrix} 4 & 6 \\ 7 & 9 \end{vmatrix} + 3 \begin{vmatrix} 4 & 5 \\ 7 & 8 \end{vmatrix}$$

The proof again follows from the definition of a determinant. In the determinant $|a_{i,j}|$, all the products that involve $a_{1,1}$ occur in the term $a_{1,1}A_{1,1}$ with proper sign since $A_{1,1}$ has all the products of one element from each row and column, excluding the first row and column. The term $a_{1,2}A_{1,2}$ likewise contains all the terms which have $a_{1,2}$ and the $-$ sign adjusts for the one interchange necessary to form the product properly, and so forth.

This leads to the next theorem.

Theorem 7   *We may expand a determinant by the elements of the ith row*

$$\sum_{k=1}^{n} (-1)^{i+k} a_{i,k} A_{i,k} = D$$

for any $i = 1, 2, \ldots, n$.

The proof follows the lines of that for Theorem 6, or else can be based directly on Theorem 6.

By the second row

$$\begin{vmatrix} 1 & 2 & 3 \\ 4 & 5 & 6 \\ 7 & 8 & 9 \end{vmatrix} = -4 \begin{vmatrix} 2 & 3 \\ 8 & 9 \end{vmatrix} + 5 \begin{vmatrix} 1 & 3 \\ 7 & 9 \end{vmatrix} - 6 \begin{vmatrix} 1 & 2 \\ 7 & 8 \end{vmatrix}$$

Theorem 8   *The alternating sum of products of elements from any row by the minors from any other row of the same determinant is zero.*

The proof follows from the observation that this is equivalent to the expansion of a determinant with two rows which are the same.

Theorems 7 and 8 can be combined into a single formula

$$\sum_{k=1}^{n} (-1)^{j+k} a_{i,k} A_{j,k} = D\delta_{i,j}$$

where $\delta_{i,j}$ is the Kronecker delta function.

Kronecker's delta function

$$\delta_{i,j} = \begin{cases} 0, & i \neq j \\ 1, & i = j \end{cases}$$

Theorem 9 (Cramer's rule)   *The solution of the system of equations*

$$a_{1,1}x_1 + a_{1,2}x_2 + \cdots + a_{1,n}x_n = b_1$$
$$a_{1,2}x_1 + a_{2,2}x_2 + \cdots + a_{2,n}x_n = b_2$$
$$\cdots\cdots\cdots\cdots\cdots\cdots\cdots\cdots\cdots\cdots\cdots\cdots\cdots\cdots\cdots$$
$$a_{n,1}x_1 + a_{n,2}x_2 + \cdots + a_{n,n}x_n = b_n$$

is given by

$$x_j D = \sum_{i=1}^{n} b_i A_{i,j} \qquad j = 1, 2, \ldots, n$$

The right-hand side is simply a column expansion of the original determinant **except** that the jth column has been replaced by the b_i's. To prove the theorem, we multiply the ith row by $A_{i,j}$ and add all the equations to get

$$x_i \sum_i a_{i,1}A_{i,j} + x_2 \sum a_{i,2}A_{i,j} + \cdots + x_j \sum a_{i,j}A_{i,j} + \cdots$$
$$+ x_n \sum a_{i,n}A_{i,j} = \sum b_i A_{i,j}$$

which by Theorems 7 and 8 is

$$D(x_1\delta_{1,j} + x_2\delta_{2,j} + \cdots + x_j\delta_{j,j} + \cdots + x_n\delta_{n,j}) = x_j D = \sum b_i A_{i,j}$$

Definition   *If all the $b_i = 0$, then the system of equations is said to be **homogeneous**.*

Corollary 1   *If $D \neq 0$, then the solution of the homogeneous system is*

$$x_1 = x_2 = \cdots = x_n = 0$$

That this is a solution is obvious. Suppose that there was also some other solution, and let $x_j \neq 0$. By Theorem 9 we are led to a contradiction since all the b_j are zero; hence the solution is unique.

Corollary 2   *If $D \neq 0$, then the system of equations has a unique solution.*

If there were two different solutions, then their difference would satisfy the corresponding homogeneous system which by Corollary 1 has only the trivial $x_i = 0$ solution.

5A.4 The case $D = 0$

If the determinant of a system of equations is zero, then there are a great many things that can be the cause of this, the most probable being that the problem is formulated incorrectly. In view of this remark we will give only a brief summary of the case $D = 0$.

First, consider only three variables so that each equation can be viewed as a plane in the three-dimensional Euclidean space.

If $D \neq 0$, then the three planes intersect in a single point.

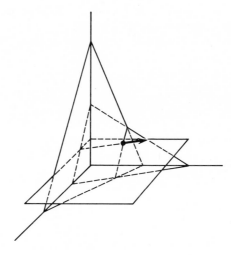

If $D = 0$, then one of the following can happen:

[1]   Two planes intersect in a line, and this line is parallel to the third plane (inconsistent equations; no solution is possible).
[2]   The three planes intersect in a common line (degenerate; many solutions are possible).
[3]   Two planes coincide and the third is parallel (inconsistent).
[4]   Two planes coincide and the third intersects the common plane (degenerate; many solutions are possible).
[5]   All three planes are parallel and distinct (inconsistent).
[6]   The three planes coincide (degenerate).

To treat the general case of n equations, where the drawings require an n-dimensional space and the complications are much more involved, we need to introduce two new ideas.

First is the idea of **rank**, which is simply the **order** (size) of the largest subdeterminant (formed by omitting various rows and columns) that is not zero.

Second is the idea of an **augmented matrix**, which is the matrix $(a_{i,j})$ with the column of b_j adjoined on the right, written as $(a_{i,j}, b_j)$; thus, the matrix has n rows and $(n + 1)$ columns. The idea of the rank of the augmented matrix is the same (except, of course, one more column than row must be omitted in forming the square subdeterminants). In Sec. 5.3 the main theorem is given.

APPENDIX: MATRICES

5B.1   Basic operations

A matrix is a rectangular array of elements $a_{i,j}$; thus,

$$A = a_{i,j} = \begin{pmatrix} a_{1,1} & a_{1,2} & \cdots & a_{1,n} \\ a_{2,1} & a_{2,2} & \cdots & a_{2,n} \\ \cdots\cdots\cdots\cdots\cdots\cdots\cdots \\ a_{m,1} & a_{m,2} & \cdots & a_{m,n} \end{pmatrix}$$

is a matrix of size $m \times n$ (m by n). Two matrices are equal if, and only if, all their corresponding elements are the same. A matrix is said to be **symmetrical** if $a_{i,j} = a_{j,i}$ for all i and j (which requires $m = n$).

Symmetrical matrix

$$\begin{pmatrix} a & b & c \\ b & d & e \\ c & e & f \end{pmatrix}$$

The sum and difference of two matrices A and B are defined as

$$A \pm B = C$$

with

$$c_{i,j} = a_{i,j} \pm b_{i,j}$$

(where, of course, A and B are of the same size).

It is clear from the definition that matrix addition is associative, that is,

$$A + (B + C) = (A + B) + C$$

and that the zero matrix O, all of whose elements are zero, plays the role of zero in the arithmetic of matrices.

$$O = \begin{pmatrix} O & O & O \\ O & O & O \\ O & O & O \end{pmatrix}$$

There are two types of multiplication to be considered. The first is multiplication by a constant k (often called "scalar" multiplication), which is defined by

$$kA = k(a_{i,j}) = (ka_{i,j})$$

In scalar multiplication **every** element of the matrix A is multiplied by the constant k.

$$k \begin{pmatrix} 1 & 2 & 3 \\ 4 & 5 & 6 \\ 7 & 8 & 9 \end{pmatrix} = \begin{pmatrix} k & 2k & 3k \\ 4k & 5k & 6k \\ 7k & 8k & 9k \end{pmatrix}$$

Since this definition differs from the corresponding one for the multiplication of a determinant by a constant k, the beginner is apt to have an uneasy feeling that inconsistencies are likely to arise and give trouble. These fears are groundless because in practice this confusion almost never occurs.

The second kind of multiplication is that for two matrices

$$\mathbf{A} \cdot \mathbf{B} = C$$

where the $c_{i,j}$ are defined by

$$c_{i,j} = \sum_{k=1}^{n} a_{i,k} b_{k,j}$$

Rule   *"Row by column"*

$$\begin{pmatrix} row \\ \rightarrow \end{pmatrix} \quad \begin{pmatrix} column \\ \downarrow \end{pmatrix}$$

Thus the elements of a row in A are term by term multiplied by the corresponding elements in a column of B and the products are summed to get the corresponding element of the product C. This definition **requires** that the number of columns of A is the same as the number of rows of B. Thus, C has the dimension of the number of rows of A and the number of columns of B. For example,

$$(a_{1,1} \quad a_{1,2} \quad \cdots \quad a_{1,n}) \begin{pmatrix} b_{1,1} \\ b_{2,1} \\ \vdots \\ b_{n,1} \end{pmatrix} = (c_{1,1})$$

has only one term where

$$c_{1,1} = \sum_{k=1}^{n} a_{i,k} b_{k,1}$$

Clearly, multiplication of two matrices is not commutative since in the above case for the product BA the elements of C are

$$\begin{pmatrix} b_{1,1} \\ b_{2,1} \\ \vdots \\ b_{n,1} \end{pmatrix} (a_{1,1} a_{1,2} \cdots a_{1,n}) = (c_{i,j}) = (b_{i,1} a_{1,j})$$

Even for square matrices the two products need not be the same since there is no reason to expect that

$$c_{i,j} = \sum_{k=1}^{m} a_{i,k} b_{k,j}$$

and

$$\overline{c_{i,j}} = \sum_{k=1}^{n} b_{i,k} a_{k,j}$$

are the same.

The identity matrix for multiplication is

$$I = \begin{pmatrix} 1 & 0 & 0 & \cdots \\ 0 & 1 & 0 & \cdots \\ 0 & 0 & 1 & \cdots \\ \cdots\cdots\cdots\cdots\cdots \end{pmatrix}$$

It is easy to see that

$$IA = AI = A$$

Diagonal matrices are more general than the identity matrix in the sense that they may have any values down the main diagonal while still having zero elements elsewhere. Evidently, the effect of multiplying on the left (right) by a diagonal matrix is to multiply the rows (columns) by the diagonal elements.

Diagonal matrix

$$\begin{pmatrix} d_1 & 0 & 0 & \cdots \\ 0 & d_2 & 0 & \cdots \\ 0 & 0 & d_3 & \cdots \\ \cdots\cdots\cdots\cdots\cdots \end{pmatrix}$$

The associative law of multiplication

$$(AB)C = A(BC)$$

however, still holds as can be seen from interchanging the summation processes in the general term of the triple product.

$$\sum_k \left(\sum_j a_{i,j} b_{j,k} \right) c_{k,l} = \sum_j a_{i,j} \left(\sum_k b_{j,k} c_{k,l} \right)$$

The definition of a product of two matrices allows us to represent a system of simultaneous linear equations,

$$\begin{aligned}
a_{1,1}x_1 &+ a_{1,2}x_2 + \cdots + a_{1,n}x_n = b_1 \\
a_{2,1}x_1 &+ a_{2,2}x_2 + \cdots + a_{2,n}x_n = b_2 \\
&\cdots\cdots\cdots\cdots\cdots\cdots\cdots\cdots\cdots\cdots\cdots\cdots \\
a_{n,1}x_1 &+ a_{n,2}x_2 + \cdots + a_{n,n}x_n = b_n
\end{aligned}$$

as a matrix product

$$Ax = b$$

where A is a square matrix $n \times n$, and x and b are column matrices of size $n \times 1$.

The definition of matrix multiplication leads immediately to the question whether

$$|A|\,|B| = |C| \quad \text{if} \quad AB = C$$

It is not obvious that this will be true, though the student is apt subconsciously to assume that it will be. One rather direct proof goes as follows. Consider the large matrix in which we have written smaller matrices as the elements to be found in the region indicated.

$$\begin{pmatrix} A & 0 \\ -I & B \end{pmatrix}$$

The determinant, by the basic definition, will be the product of all possible products (of the usual type) of A by one of the products of the usual type of B regardless of what is written in the lower left-hand corner. Thus, clearly,

$$\begin{vmatrix} A & 0 \\ -I & B \end{vmatrix} = |A|\,|B|$$

Writing the whole matrix out in a little detail, we have

$$\begin{pmatrix} a_{1,1} & a_{1,2} & \cdots & 0 & 0 & \cdots \\ a_{2,1} & a_{2,2} & \cdots & 0 & 0 & \cdots \\ & & \cdots\cdots\cdots & & & \\ -1 & 0 & \cdots & b_{1,1} & b_{1,2} & \cdots \\ 0 & -1 & \cdots & b_{2,1} & b_{2,2} & \cdots \\ & & \cdots\cdots\cdots & & & \end{pmatrix}$$

We are going to use the 1s in the lower left to eliminate the a's in the upper left. When we examine how we eliminate the ith row of A, we shall need to use for the multipliers of the lower rows the numbers $a_{i,1}, a_{i,2}, \ldots$, and these will multiply the corresponding b's in the jth column, $b_{1,j}, b_{2,j}, \ldots$; thus, we shall find in the upper right-hand part exactly

$$\sum_{k=1}^{n} a_{i,k} b_{k,j} \equiv c_{i,j}$$

so that we now have, after the elimination of all the a's in the upper left,

$$\begin{pmatrix} 0 & C \\ -I & B \end{pmatrix}$$

We now interchange the rows of the upper half with the rows of the lower half; each interchange of course produces a sign change which we absorb by changing the sign of the 1 in the identity matrix. We have, therefore,

$$\begin{pmatrix} I & B \\ 0 & C \end{pmatrix}$$

Expanding this by the basic definition, we find that we have the determinant of C. Therefore, the definition of matrix multiplication is "consistent with" the definition of the value of a determinant, and we have

$$|A|\,|B| = |C|$$

INTERPOLATION AND ROUNDOFF ESTIMATION

6

6.1 Linear interpolation

Interpolation is usually first introduced in connection with the study of logarithms and trigonometry. The tables of logarithms and the trigonometric functions are generally arranged so that **linear interpolation** will give sufficient accuracy (for almost all values).

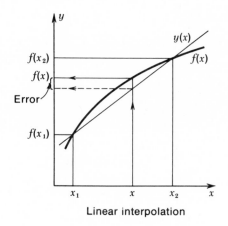

Linear interpolation

The usual table of a function $f(x)$ gives the values for $f(x_i)$ to a fixed number of decimal places for a sequence of equally spaced values of x_i. From these approximate values of $f(x_i)$ we are asked to **estimate** the value of $f(x)$ for some value x that is not in the table. This is often called "reading between the lines."

The process of doing linear interpolation is simple. Given a value x, we search in the table for a pair of values, which we shall call x_1 and x_2, such that

$$x_1 \leq x < x_2$$

If $x_1 \neq x$, we then **assume** that between the two values x_1 and x_2 the function $f(x)$ can be approximated by a straight line through the two points. This line is given by

$$y(x) = f(x_1) + \frac{f(x_2) - f(x_1)}{x_2 - x_1} \ (x - x_1)$$

$$= \frac{f(x_1) \ [x_2 - x] + f(x_2) \ [x - x_1]}{x_2 - x_1}$$

and is the desired formula for estimating the value of $f(x)$ from the given value of x.

When we are at the end of a table and wish to estimate a value beyond the end, we use the same formula but call it **extrapolation**. As we shall later see, extrapolation is usually much more dangerous than is interpolation.

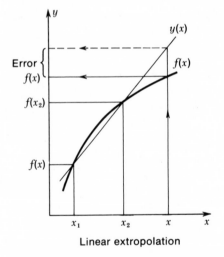

Linear extropolation

If we are given the value of the function and are asked to find the corresponding value of x, then the process is called **inverse interpolation**.

For inverse interpolation the same straight line is used, but the equation is rearranged into the more convenient form

$$x = x_1 + (x_2 - x_1) \left[\frac{f(x) - f(x_1)}{f(x_2) - f(x_1)} \right]$$

$$= \frac{x_1 \ [f(x_2) - f(x)] + x_2 \ [f(x) - f(x_1)]}{f(x_2) - f(x_1)}$$

where $f(x)$ is the given value of the function which we wrote before as $y(x)$. Note the symmetry of the equations for the two types of interpolation.

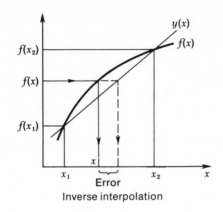

Inverse interpolation

PROBLEMS 6.1

1 Derive the inverse interpolation formula.
2 Linearly interpolate for sin 30° from the values sin 0° = 0 and sin 45° = 0.707.
3 Inversely interpolate for sin x = 0.5 from sin 0° = 0 and sin 45° = 0.707.

6.2 Analytic substitution

The process of linear interpolation is an example of **analytic substitution**. In place of the function $f(x)$, which we cannot handle, we substitute a straight line and use that **as if it were** the function $f(x)$. For linear interpolation the particular straight line was chosen to pass through the two end points of the interval, where we knew the values of the function. In Chap. 2 when we treated Newton's method, we also used a straight line in place of the function whose zero we were seeking, but we used the value of the function and the slope at the same point, rather than the value of the function at two distinct points.

Newton's method

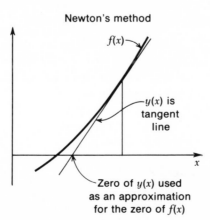

The idea of analytic substitution is central to much of numerical analysis. Repeatedly, when we have a function which we cannot handle, we replace it by some other analytic expression which we can handle and operate on the new expression **as if it were** the original function. Two steps are involved in analytic substitution. First, what class of functions shall we use? Up to now we have discussed the class of straight lines. Second, how shall we select the particular member of the class? Here we have tried both using two points to determine the straight line and using a point plus a slope at the same point to determine the line.

We shall later examine two other criteria for picking the approximating function: least squares in Chap. 10 and the Chebyshev, or minimax, criterion in Chap. 13.

Classical numerical analysis generally uses a class of polynomials up to some fixed degree, and we shall do this most of the time. However, in Chap. 12 we shall consider the class of sines and cosines as our approximating functions.

Polynomials

$$y(x) = a_0 + a_1 x + \cdots + a_N x^N$$

Fourier series

$$y(x) = a_0 + a_1 \cos x + a_2 \cos 2x + b_1 \sin x + b_2 \sin 2x$$

PROBLEM 6.2

1 Describe the "analytic substitution" used in the:
 a False position method.
 b Secant method.

6.3 Polynomial approximation

The values that we are given of a function are sometimes spaced so far apart that linear interpolation is not sufficiently accurate for our purposes. In such cases classical numerical analysis uses the class of nth-order polynomials

$$P_n(x) = a_0 + a_1 x + a_2 x^2 + \cdots + a_n x^n$$

and chooses a polynomial which passes through selected samples of the function.

The simplest method to understand for finding a polynomial is **the method of undetermined coefficients**, which we shall use repeatedly. This method assumes the **form** of the answer with arbitrary (undetermined) coefficients written in it. Then the conditions, such as passing through the various points, are applied to determine the arbitrary coefficients.

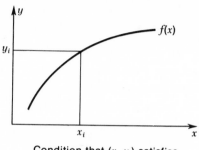

Condition that (x_i, y_i) satisfies
the expression is $y_i = f(x_i)$

In the particular case of polynomial interpolation in a table of values of a function $y(x)$, the condition that the polynomial pass exactly through the point (x_i, y_i) is that

$$P_n(x_i) = y_i = a_0 + a_1 x_i + \cdots + a_n x_i^{\,n}$$

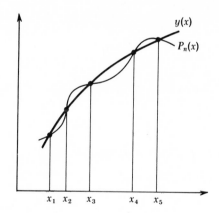

If there are as many points, equally spaced or not, as there are undetermined coefficients (parameters) a_j, then we must have $n + 1$ points and are led to $n + 1$ linear equations in the unknowns a_j:

$$y_i = \sum_{j=0}^{n} a_j x_i^{\,j} \qquad (i = 1, 2, \ldots, n + 1)$$

The determinant of the unknown coefficients a_j is the famous Vandermonde determinant

$$D = |x_i^j| = \begin{vmatrix} 1 & x_1 & x_1^2 & \cdots & x_1^n \\ 1 & x_2 & x_2^2 & \cdots & x_2^n \\ \hdotsfor{5} \\ 1 & x_{n+1} & x_{n+1}^2 & \cdots & x_{n+1}^n \end{vmatrix}$$

which cannot be zero if $x_i \neq x_j$ as we shall prove in the next section. We can therefore, at least in principle, determine the unique interpolating polynomial. Interpolation is, then, merely a matter of evaluating this polynomial at the desired point(s).

Example   *Find cubic $P_3(x)$ such that*

$$P_3(0) = 0$$
$$P_3'(0) = 1$$
$$P_3(1) = 1$$
$$P_3'(1) = 0$$

Solution

P_3	$=$	a_0	$+$	$a_1 x$	$+$	$a_2 x^2$	$+$	$a_3 x^3$	
P_3'	$=$			a_1	$+$	$2a_2 x$	$+$	$3a_3 x^2$	
$P_3(0)$	$=$	a_0							$= 0$
$P_3(1)$	$=$	a_0	$+$	a_1	$+$	a_2	$+$	a_3	$= 1$
$P_3'(0)$	$=$			a_1					$= 1$
$P_3'(1)$	$=$			a_1	$+$	$2a_2$	$+$	$3a_3$	$= 0$

and $a_0 = 0, \quad a_1 = 1$

$$\begin{cases} a_2 & + & a_3 & = 0 \\ 2a_2 & + & 3a_3 & = -1 \end{cases} \longrightarrow \begin{cases} a_2 = 1 \\ a_3 = -1 \end{cases}$$

Therefore, $P_3(x) = x + x^2 - x^3$

PROBLEMS 6.3

1. Construct the interpolating quadratic through $[a, f(a)]$, $[(a + h, f(a + h)]$, and $[a + 2h, f(a + 2h)]$.
2. Construct an interpolating cubic through the points (0,1), (1,2), (2,2), and (3,3).
3. Construct an interpolating cubic through the points $(-1,1)$, (0,0), (1,1), and (2,4).

6.4   The Vandermonde determinant

To show that if $x_i \neq x_j$, then the Vandermonde determinant

$$D = \begin{vmatrix} 1 & x_1 & x_1^2 & \cdots & x_1^n \\ 1 & x_2 & x_2^2 & \cdots & x_2^n \\ \multicolumn{5}{c}{\cdots\cdots\cdots\cdots\cdots\cdots\cdots} \\ 1 & x_{n+1} & x_{n+1}^2 & \cdots & x_{n+1}^n \end{vmatrix}$$

is not zero, we ignore where it came from and regard it as a function of the variables x_i ($i = 1, 2, \ldots, n + 1$),

$$D = D(x_i, x_2, \ldots, x_{n+1})$$

First, D is clearly a polynomial in the x_i. What is the degree of this polynomial regarded as a function of all the x_i? Each term in the expansion of the determinant has an element from each row and each column. The element from the first column has 0 degree, the element from the second column has first degree, the element from the next column has second degree, and so forth, and in total the degree of the product is

$$1 + 2 + 3 + \cdots + n = \frac{n(n + 1)}{2}$$

Degree of this term is
$0 + 1 + 2 + \cdots + n$
as is the degree of
every other term

Next, it is evident that if $x_1 = x_2$, then two rows of the determinant are the same and $D = 0$. Similarly, if $x_1 = x_3$, then $D = 0$; if $x_1 = x_4$, then $D = 0$; etcetera, until $x_1 = x_{n+1}$. Thus, D must have all the factors

$$\prod_{i=2}^{n+1} (x_i - x_1) = (x_2 - x_1)(x_3 - x_1) \cdots (x_{n+1} - x_1)$$

Furthermore, $D = 0$ if **any** $x_i = x_j$, and D must have all the factors

$$\left. \begin{array}{l} (x_2 - x_1)\ (x_3 - x_1)\ \cdots\ (x_{n+1} - x_1) \\ (x_3 - x_2)\ \cdots\ (x_{n+1} - x_2) \\ \cdots\cdots\cdots\cdots\cdots\cdots\cdots\cdots\cdots\cdots\cdots\cdots\cdots \\ (x_{n+1} - x_n) \end{array} \right] = \prod_{i>j=1}^{n+1} (x_i - x_j)$$

The degree of this polynomial in the x_i's is found by simply counting the number of linear factors

$$n + (n-1) + \cdots + 1 = \frac{n(n+1)}{2}$$

Thus, except for a multiplicative constant, the polynomial we have just constructed is $D(x_1, x_2, \ldots, x_{n+1})$.

To find this constant we need only compare the same term in the expansions of the two expressions, the determinant form and the product form. The main diagonal term of the determinant is

$$1\ x_2\ x_3^2\ \cdots\ x_{n+1}^n$$

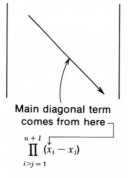

Main diagonal term
comes from here

$$\prod_{i>j=1}^{n+1} (x_i - x_j)$$

which has a coefficient of $+1$. When we expand the product form and search for this same term, we find that it comes from the terms on the left sides of the parentheses and hence also has a coefficient of $+1$. We conclude, therefore, that

$$D = \prod_{i>j=1}^{n+1} (x_i - x_j)$$

and $D \neq 0$ for $x_i \neq x_j\ (i \neq j)$.

6.5 The error term

Having found the approximating polynomial, we naturally ask how much error we make in the analytic substitution process when we use this polynomial $P_n(x)$ in place of the original function $y(x)$. We therefore examine the difference

$$y(x) - P_n(x)$$

This difference is zero at each $x = x_i$; hence, we can write

$$y(x) - P_n(x) = (x - x_1)(x - x_2) \cdots (x - x_{n+1})K(x)$$

where $K(x)$ is some function of x. For any value of x, say $\bar{x}$ (we are not using complex numbers so this is not the conjugate), we can write

$$y(\bar{x}) = P_n(\bar{x}) = (\bar{x} - x_1)(\bar{x} - x_2) \cdots (\bar{x} - x_n)K(\bar{x})$$

Consider, now, the expression

$$\phi(x) \equiv y(x) - P_n(x) - (x - x_1)(x - x_2) \cdots (x - x_{n+1})K(\bar{x})$$

where the $\bar{x}$ occurs only in the $K(\bar{x})$ term. For this expression we know that

$$\phi(x) = 0 \quad \text{for} \quad x = x_1, x_2, \ldots, x_{n+1} \quad \text{and} \quad \bar{x}$$

If we differentiate this expression once, we get, because of the mean-value theorem,

$$\phi'(x) = 0$$

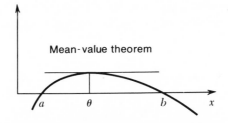

Mean-value theorem

There is a θ where tangent is horizontal

for at least $n + 1$ values of x. If we differentiate again, by the same reasoning we have

$$\phi''(x) = 0$$

for at least n values of x.

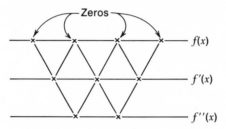

Continuing in this manner, we come finally to

$$\phi^{n+1}(x) = 0$$

for at least one value of x, say, $x = x^*$, **provided** the function $y(x)$ has at least $n + 1$ derivatives. When we actually carry out this differentiation with respect to x on the term

$$(x - x_1)(x - x_2) \cdots (x - x_{n+1})K(\bar{x})$$

we find that we have

$$\phi^{n+1}(x^*) = y^{n+1}(x^*) - (n + 1)!K(\bar{x}) = 0$$

The expression for $K(\bar{x})$ is therefore

$$K(\bar{x}) = \frac{y^{n+1}(x^*)}{(n + 1)!}$$

and putting this back in the original expression, we have finally

$$y(x) = P_n(x) + \frac{(x - x_1)(x - x_2) \cdots (x - x_{n+1})}{(n + 1)!} y^{n+1}(x^*)$$

where in place of writing the arbitrary value $\bar{x}$, we have simply written x. The x^* lies inside the interval containing all the x_i **and** x.

Thus, we have the error term for the polynomial approximation, which uses $n + 1$ points of the function to match the polynomial. This error term is often called the **truncation error** of the approximation.

This formula gives exactly the error in the approximation, but unfortunately it has a pair of weaknesses. First, the value x° to be used is unknown. Second, except for comparatively few functions it is not practical to compute an $(n + 1)$st derivative if n is at all large.

How shall we assess these faults? In the first place, it is unreasonable to expect to get a useful formula for the exact error, that is, to know the correct value of a function when all we were given were $n + 1$ samples of the function. In the second place, while the exact error cannot be found practically, the formula for the exact error often provides very practical bounds on the error or at least a guide in choosing among competing formulas.

PROBLEM 6.5

1 Write out the error term for quadratic and cubic interpolation.

6.6 Some remarks on the error term

There are many different forms for polynomial interpolation, but owing to the uniqueness of the interpolating polynomial, the answer is uniquely determined (except for roundoff differences due to the way it is computed) once the sample points x_i are chosen to exactly match the function.

The interpolating polynomial of degree n through n + 1 points is unique no matter how we find it, because if there were two polynomials $P_1(x)$ and $P_2(x)$, then their difference

$$\phi(x) = P_1(x) - P_2(x)$$

would be of degree n (at most) and would have n + 1 zeros, which makes

$$\phi(x) \equiv 0$$

Thus we shall simply ignore the many different formulas, which are often very useful for some particular situation; they tend to save time at the expense of human memory.

From a careful study of the error term it should become clear that insofar as we can, we should take the sample points clustered around the value x where we are going to interpolate so that the one term we have under our control

$$(x - x_1)(x - x_2) \cdots (x - x_{n+1})$$

is as small as we can make it.

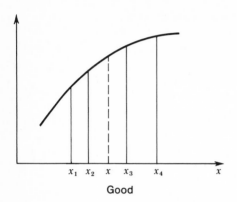

Good

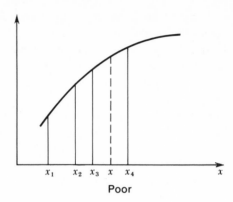

Poor

It should also be clear that when we extrapolate (use an x outside the range of the sample points x_i), then this factor is apt to be large.

Although we have referred to tables as the source of our numbers, it should be realized that the various answers from a computer are often associated with a sequence of parameter values and they form a table. Therefore, what we have been examining is very relevant to much of modern computing.

A TABLE OF
VALUES OF $f(x)$

x	$f(x)$
1	—
2	—
3	—
4	—
⋮	⋮

For example, when a large number of closely spaced values are wanted, it may be worthwhile computing the values at a much wider spacing and then using an interpolating process to get the missing values. Such would be the case if each value requires a great deal of machine time to compute; otherwise it is probably not worth thinking about.

6.7 Equally spaced data

Equally spaced values of the independent variable are so frequent and the simplifications in the theory are so great that they are worth special treatment.

We first introduce the notation (familiar from the calculus) of the forward difference

$$\Delta y_i = y_{i+1} - y_i \quad (i = 0, 1, \ldots, n)$$
$$= y[a + (i + 1)h] - y(a + ih)$$

where a is an arbitrary constant that indicates where the origin of the data

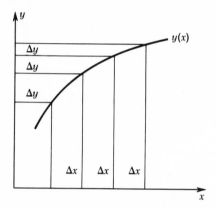

is, and h is the spacing of the data $\Delta x = h = x_{i+1} - x_i =$ some constant $\neq 0$

The substitution

$$x = a + h\bar{x}$$

causes $\bar{x}_i$ to have the values 0, 1, 2, ..., n and simplifies the notation greatly. We shall therefore frequently act as if we had this unit spacing with the origin at zero.

As in the case of the derivative operator d/dx in the calculus, we can apply the Δ operator again and again:

$$\frac{d}{dx}\frac{dy}{dx} = \frac{d^2y}{dx^2}$$

$$\Delta(\Delta y_i) = \Delta^2 y_i = \Delta y_{i+1} - \Delta y_i$$
$$= y_{i+2} - 2y_{i+1} + y_i$$

$$\Delta^3 y_i = \Delta^2 y_{i+1} - \Delta^2 y_i$$
$$= y_{i+3} - 3y_{i+2} + 3y_{i+1} - y_i$$

and so forth. It is probably best to picture the differences as if they were in the difference table form

$$
\begin{array}{cccc}
y_0 \\
& > \Delta y_0 \\
y_1 & & > \Delta^2 y_0 \\
& > \Delta y_1 & & > \Delta^3 y_0 \\
y_2 & & > \Delta^2 y_1 & \quad . \\
& > \Delta y_2 & \quad . & \quad . \\
y_3 & \quad . & \quad . & \quad . \\
\quad . & \quad . & \quad . & \quad . \\
\quad . & \quad . & \quad . & \quad . \\
\quad . \\
\end{array}
$$

....

though, of course, they are not stored that way in a machine.

A DIFFERENCE TABLE

x	y	Δy	$\Delta^2 y$	$\Delta^3 y$
0	0			
		1		
1	1		6	
		7		6
2	8		12	
		19		6
3	27		18	
		27		6
4	64		24	
		61		
5	125			

It is convenient to introduce a second operator E, called the **shift operator**, which is defined by

$$E\{y(x)\} = y(x + h)$$

Obviously,

$$E^k\{y(x)\} = y(x + kh)$$

and

$$E^0\{y(x)\} = y(x)$$

The two operators Δ and E are connected by the relation

$$\Delta y_i = y_{i+1} - y_i = Ey_i - 1y_i = (E - 1)y_i$$

for an arbitrary function y. Thus we have the abstract **operator equations**

$$E - 1 = \Delta$$
$$E = \Delta + 1$$

We can use these operators to compute the higher differences directly in terms of the function values y_i

$$\Delta^k = (E - 1)^k = E^k - kE^{k-1} + \frac{k(k-1)}{2} E^{k-2} - \cdots$$
$$= E^k - C(k,1) E^{k-1} + C(k,2) E^{k-2} - \cdots$$

where we have used the convenient notation $C(n,k)$ for the kth binomial coefficient of order n,

$$C(n,k) = \frac{n!}{k!(n-k)!}$$

instead of the currently usual notation $\binom{n}{k}$.

Case of $k = 3$

$$\Delta^3 y_0 = y_3 - 3y_2 + 3y_1 - y_0$$

The expression

$$E^k = (1 + \Delta)^k$$

leads to

$$y(x + k) = y(x) + k\,\Delta y(x) + \frac{k(k-1)}{2}\,\Delta^2 y(x) + \cdots + \Delta^k y(x)$$
$$= \sum_{i=0}^{k} C(k,i)\,\Delta^i y(x)$$

Case of $k = 3$

$$y(x + 3) = y(x) + 3\Delta y(x)$$
$$+ 3\Delta^2 y(x)$$
$$+ \Delta^3 y(x)$$

Both of these equations are simply identities and contain nothing essentially new.

PROBLEMS 6.7

1 Discuss the systematic subtabulation of a table of values when we wish to insert four equally spaced values between each equally spaced original table entry by using cubic interpolation. (Ignore the end intervals.)
2 Write out the formula for the fifth difference expression in terms of the function values.

6.8 Differences of Polynomials

Just as one finds that the $(k + 1)$st derivative of a polynomial of degree k (or less) is zero, we need to prove the corresponding important

$$\frac{d^{k+1}}{dx^{k+1}} P_k(x) = 0$$

Theorem   *The kth difference of a polynomial of degree k*

$$y(x) = a_0 + a_1 x + \cdots + a_k x^k$$

is identically a constant $a_k k! h^k$ and the $(k + 1)$st difference is identically zero.

Note that x is **any** value you choose, not necessarily an integer. The proof is easily obtained from the following

$$\Delta^{k+1} P_k(x) = 0$$

Lemma   *If $y(x)$ is a polynomial of degree k (exactly), then $\Delta y(x)$ is a polynomial of degree $k - 1$ (and no lower).*

Proof   To prove the lemma, we start with the special polynomial $y = x^n$

$$\begin{aligned}
\Delta y &= (x + h)^n - x^n \\
&= x^n + nhx^{n-1} + \cdots + h^n - x^n \\
&= nhx^{n-1} + \cdots
\end{aligned}$$

Thus Δ drops each single power to a polynomial of lower degree; hence, since Δ is a linear operator,

$$\Delta \sum_{0=1}^{k} a_i x^i = \sum_{i=1}^{k} a_i \Delta x^i$$

is a polynomial of degree $k - 1$ with leading coefficient equal to

$$a_k kh$$

If now we apply the lemma repeatedly, we get, after k times,

$$a_k k! h^k$$

and after $k + 1$ times, exactly 0.

The reason that this theorem is so important (in classical numerical analysis) is that we tend to tabulate our results at a small enough interval so that the function may be quite closely approximated **locally** by a polynomial of moderate degree. If as we take higher differences, it turns out that the differences of our (equally spaced) results do not get small rapidly, then usually we compute the results at a spacing half as large.

TABLES OF THE
FUNCTION $Y = X^2$ AT
DIFFERENT SPACINGS

x	$y = x^2$	Δy	$\Delta^2 y$	x	y	Δy	$\Delta^2 y$	x	y	Δy	$\Delta^2 y$
0	0			0	0			0	0		
		4				1				$\frac{1}{4}$	
2	4		8	1	1		2	$\frac{1}{2}$	$\frac{1}{4}$		$\frac{1}{2}$
		12				3				$\frac{3}{4}$	
4	16		8	2	4		2	1	1		$\frac{1}{2}$
		20				5				$\frac{5}{4}$	
6	36		8	3	9		2	$\frac{3}{2}$	$\frac{9}{4}$		$\frac{1}{2}$
		28				7				$\frac{7}{4}$	
8	64		8	4	16		2	2	4		
		36				9					
10	100			5	25						

The effect of halving the spacing in x is to divide (approximately) the first differences by 2, the second differences by 4, the third differences by 8, and so forth. Thus, by choosing a sufficiently small spacing, we should in most practical cases be able to get the difference table to have small values.

This remark must be tempered by the fact that we compute with the numbers that are actually in the machine and the theorem is a typical mathematical proof concerned with infinitely precise numbers. Thus, we turn in Sec. 6.10 to the question of roundoff.

PROBLEMS   6.8

1 Find the first and second differences (for $h = 1$) of

$$y = 2x^2 - 6x - 7$$

2 Show how to construct successive values of a polynomial from the starting differences.

6.9 An example of table extrapolation with polynomials

The manufacturing costs of a certain computer component were estimated to be

Year	Cost
1968	63
1969	52
1970	43
1971	38

How are we to judge this, and in particular how can we extend the table?

	C	ΔC	$\Delta^2 C$	$\Delta^3 C$
1968	63			
1969	52	-11	2	
1970	43	-9	4	2
1971	38	-5		

If we assume that it is a cubic, then the third differences are constant and we can fill in the last column with 2s. From this we can calculate in turn $\Delta^2 C$, ΔC, and C for each year.

Year	Cost	Δ	Δ²	Δ³
1968	63			
1969	52	−11	2	
1970	43	−9	4	2
1971	38	−5	6	2
1972	39	1	8	2
1973	48	9	10	2
1974	67	19	12	2
1975	98	31		

The manufacturing costs are rising rapidly!
   But we note that changing the last two given estimated costs to

1970	44
1971	39

(which is a small change) lets us fit a quadratic:

Year	Cost	Δ	Δ²	Δ³
1968	63			
1969	52	−11	3	
1970	44	−8	3	0
1971	39	−5	3	
1972	37	−2	3	
1973	38	1	3	
1974	42	4	3	
1975	49	7		

whose extrapolated value at 1975 is one-half that of the cubic!
   Polynomials are simply not suited for extrapolation, but if you insist
on using them, then the lower-order ones seem to be safer. Note that
both have a minimum around 1972 and that the quadratic provides the
more believable estimate, but is hardly a reliable prediction.
   Further examination suggests adjusting the 1971 value **only** to fit a
quadratic

Year	Cost	ΔC	$\Delta^2 C$
1968	63		
		−11	
1969	52		2
		−9	
1970	43		2
		−7	
1971	36		2
		−5	
1972	31		2
		−3	
1973	28		2
		−1	
1974	27		2
		+1	
1975	28		

Here the adjustment of the 1971 value from 38 to 36 (not a large change in view of the uncertainty of the estimates) gives a still more believable set of costs, but by this time presumably our confidence has been destroyed and we suspect we could produce almost anything we wanted if we fooled around long enough.

As an alternative approach, consider that perhaps it is the log C that is important.

	$\log C$	Δ
1968	1.799	−83
1969	1.716	−83
1970	1.633	−53
1971	1.580	

Suppose we alter the 1971 value so that the Δ is constant. We should have

$$\log C = 1.550$$
$$C = 35.5$$

We should then have the table for extrapolation.

	C	$\log C$	Δ
1968	63	1.799	
			−83
1969	52	1.716	
			−83
1970	43	1.633	
			−83
1971	35.5	1.550	
			−83
1972	. . . .	1.467	
			−83
1973	. . . .	1.384	
			−83
1974	. . . .	1.301	
			−83
1975	16.5	1.218	

Again, we could slightly modify the data.

	C	log C	Δ
1968	63	1.799	
			-83
1969	52	1.716	
			-73
1970	44	1.643	
			-63
1971	38.5	1.580	
			-53
1972	33.5	1.527	
			-43
1973	30.5	1.484	
			-33
1974	28.5	1.451	
			-23
1975	27.0	1.428	

A rather different cost in 1975.

6.10 Roundoff

Suppose first we had a function which was zero identically, and that we made a small error ε at only one point. Let us examine the difference table

y	Δy	$\Delta^2 y$	$\Delta^3 y$
0			
	0		
0		0	
	0		ε
0		ε	
	ε		-3ε
ε		-2ε	
	$-\varepsilon$		3ε
0		ε	
	0		$-\varepsilon$
0		0	
	0		
0			

We see immediately the growing triangle of ε's having binomial coefficients which grow rapidly. Thus the difference table tends to magnify small errors.

Second, consider the computed table as the product of the true values $y(x_i) = y_i$ and of random roundoff factors of size $(1 + \varepsilon_i)$. Thus, the values we have are

$$y_i(1 + \varepsilon_i) = y_i + y_i \varepsilon_i$$

When we compute the difference table of this function, we get the same result as if we had computed the differences of two tables and added them; the first table of differences being that of the original function y_i and the second table of differences being that of the round-off $y_i \varepsilon_i$. The reason that this is so is, of course, that the operator Δ is linear and the difference of a sum is the sum of the differences.

Find the error in the table.

	Δ	Δ^2	Δ^3
246			
272	-26		
	-28	2	
300		2	0
	-30		5
330		7	
	-37		-5
367		2	1
	-39	3	
406	-42		
448			

We suspect that Δ^3 shows an error. The pattern

> low
> high
> low
> high

is centered at about 330.

The 5, -5 *suggests*

$$-3\varepsilon, \ 3\varepsilon$$

The difference

$$-5 - (5) = -10 \sim 6\varepsilon$$

suggests that

$$\varepsilon = -2$$

Try 332 *in place of* 330:

		Δ	Δ^2
	246		
	272	-26	
		-28	2
	300		4
		-32	
change	332		3
		-35	
	367		4
		-39	
	406	-42	3
	448		

It looks much better!

Notice in computing the difference table that usually y_i and y_{i+1} have the same sign and therefore there is a cancellation and no loss of absolute accuracy even if the numbers do get smaller. It is only in the occasional situation where the algebraic signs of consecutive table values are different that an addition can occur and hence the possibility of a carry to the left which forces a roundoff that loses information.

PROBLEM 6.10

1 Find the error in the table

2460
2718
3004
3318
3669
4055
4482
4957

6.11 Philosophy

From the beginning we have stressed that we are using finite machines. One consequence of this is that we must recongnize that roundoff can occur and at times do us serious harm if we are not careful.

We are about to enter deeply into the domain of the infinite processes in mathematics which of necessity must be approximated by finite processes. Thus, in addition to roundoff we face on a computer a second source of trouble, truncation error.

Truncation error versus roundoff error.

*Smaller steps **usually** reduce truncation error and **may** increase roundoff error.*

Just as we recognized that we were going to use the words "small" and "large" when talking about numbers and therefore had to do some thinking about scaling to give these words meaning, so now we must try to create some kind of a theory which will enable us to separate the two effects that arise from the finiteness of the machine.

Finite number length leads to roundoff.

Finite processes lead to truncation.

The theory we shall produce rests on the two observations:

1 The higher differences of a suitably tabulated smooth function **tend** to approach zero (Sec. 6.8)
2 The higher differences due to random roundoff errors **tend** to get large (Sec. 6.10).

The weakness of the theory was carefully indicated by the word "tend" in both statements. We are not going to present a perfect, reliable, elegant theory; rather we are going to develop what might be called a "desperation theory" whose chief merit is that it is better than no theory at all. If the truncation error is too large, **usually** we can reduce it by a known multiplicative factor $\frac{1}{2}$ ^{k} by halving the interval, (Sec. 6.8), but if we are already down to the roundoff level of accuracy, we shall clearly do ourselves harm (as well as increase the computing bill) with no gain in accuracy.

Truncation error is typically of the form

$$E_k = C\,\frac{h^k y^{(k)}\,(\theta)}{k!}$$

but may be

$$E_k = C\,\frac{h^k y^{(k-1)}\,(\theta)}{(k-1)!}$$

Let us be clear about what we can expect from a roundoff theory. We could expect to state a bound on the error. If we develop this approach, we find that often the guaranteed bound is so pessimistic that we cannot afford to use it. Also it is likely to cost us dearly either in our own time or in machine time.†

We intend to develop a statistical theory for estimating roundoff. Just as an insurance company by using statistics can often make quite successful predictions about the number of deaths they will have to pay off on (although on any individual they may be very wrong in estimating his death date), so too we shall make estimates for average behavior only and shall at times make gross errors. As we said, it is a desperation theory. Epidemics of trouble may occur and invalidate our predictions.

---

†A method now being developed under the name "range arithmetic" (also called "interval arithmetic") will probably provide one solution when it becomes generally available.

6.12 The roundoff model

We will assume that our numbers y_i have typical floating point errors and appear to us as

$$y_i(1 + \varepsilon_i)$$

where y_i and ε_i are uncorrelated. We will assume that "on the average" ε_i is as likely to have one sign † as the other; thus, we assume

$$\text{Ave } \{\varepsilon_i\} = 0$$

We are, of course, using the customary statistical device of thinking of many repetitions of the same experiment (calculation) and of getting an **ensemble** of ε_i. In practice the same input, barring machine failures, gives the same result. We are only "thinking"; thus, our averages of the ε_i are taken over the **imagined** repetitions.

In addition to assuming ave $\{\epsilon_i\} = 0$, we assume that the error ε_i made in the ith function value is uncorrelated with those at other points. Uncorrelated means

$$\text{Ave } \{\varepsilon_i \varepsilon_j\} = 0 \qquad i \neq j$$

Often, especially in recursive calculations, this assumption is plainly false and we must be careful how we apply the conclusions we are going to develop. As we noted, epidemics do occur, especially on computing machines that "chop" rather than round.

6.13 Roundoff in the kth difference

Let us fix in our minds a particular position in the kth difference column of the roundoff error table of values $y_i \varepsilon_i$. The entries which are far removed from the difference entry we are looking at are not used in its calculation. As we go up the table from the bottom, we come to a first entry that is involved in the computation of $\Delta^k y_i \varepsilon_i$. The entry $y_{i+k} \varepsilon_{i+k}$ will have a coefficient $+1$. The entry $y_{i+k-1} \varepsilon_{i+k-1}$ which follows

---

†On many computers the numbers are not rounded but are "chopped" by dropping the extra digits. If positive and negative numbers of the same size are about equally likely, then this model is reasonable; otherwise it is not.

will have a coefficient $-C(k,1)$, then $y_{i+k-2}\varepsilon_{i+k-2}$ with coefficient $+C(k,2)$, and so forth. In all, we shall have the usual formula

$$\Delta = E - 1$$
$$\Delta^k = (E - 1)^k$$

$$\Delta^k\{y_i\varepsilon_i\} = \sum_{j=0}^{k} (-1)^j C(k,j)\, y_{i+k-j}\varepsilon_{i+k-j}$$

$$= \sum_{j=0}^{k} (-1)^{k-j} C(k,j)\, y_{i+j}\varepsilon_{i+j}$$

for computing the kth difference we fixed our attention on.

How big is this number? If we **bound** the number we shall get

$$|\Delta^k\{y_i\varepsilon_i\}| \leq 2^k \max_{j=0,\dots,k} (y_{i+j}\varepsilon_{i+j})$$

$$\sum_{j=0}^{k} C(k,j) = (1 + 1)^k = 2^k$$

If we calculate the average, we shall, by the nature of our assumption on the behavior of the errors ε_i, get exactly zero (the y_i are fixed and the average is over the ensemble of the ε_i).

It is much more reasonable to ask, "What is the average of the square of the kth difference? That is,

$$\text{Ave}\ \{(\Delta^k y_i\varepsilon_i)^2\}$$

since in this kind of expression there will be no cancellation of plus and minus values of $\Delta^k y_i\varepsilon_i$,

$$Variance\ \{x_i\} = \frac{1}{n}\sum_{i=1}^{n} (x_i - \bar{x})^2$$

where $\bar{x} =$ mean of the x_i. In text, the mean of the $y_i\varepsilon_i$ is zero.

all terms will contribute according to the square of their size. Big errors, being more serious than small ones, are taken more seriously in the calculation of the average. We might have tried

$$\text{Ave}\ \{|\Delta^k y_i\varepsilon_i|\}$$

but this is unmanageable theoretically, and we stick with the **variance**, as it is called in statistics books.

Substituting our expression for $\Delta^k y$, squaring out, and taking the average term by term we get, recalling the lack of correlation of the ε_i,

$$\text{Ave}\ \{(\Delta^k y_i \varepsilon_i)^2\} = \text{Ave}\ \left\{ \sum_{j=0}^{k} (-1)^j C(k,j) y_{i+j} \varepsilon_{i+j} \sum_{m=0}^{k} (-1)^m C(k,m) y_{i+m} \varepsilon_{i+m} \right\}$$

$$= \sum_{j=0}^{k} \sum_{m=0}^{k} (-1)^{j+m} C(k,j) C(k,m) y_{i+j} y_{i+m}\ \text{Ave}\ \{\varepsilon_{i+j} \varepsilon_{i+m}\}$$

$$= \sum_{j=0}^{k} C^2(k,j) y^2_{i+j}\ \text{Ave}\ \{(\varepsilon_{i+j})^2\}$$

Remember that the ε_i's being uncorrelated means

$$\text{Ave}\ \{\varepsilon_{i+j} \varepsilon_{i+m}\} = 0 \qquad (j \neq m)$$

Holding this formula in the back of our minds, let us ask what the variance is in the error terms of the orignal function values; this is by definition

$$\text{Ave}\ \{(y_i \varepsilon_i)^2\} = \sigma^2$$

We call this quantity, σ^2, the variance or "noise level" in the original table.

Now, returning to the formula we were examining and noting that

$$\text{Ave}\ \{y_i^2 \varepsilon_i^2\} = y_i^2\ \text{Ave}\ \{\varepsilon_i^2\} = \sigma^2$$

we have

$$\text{Ave}\ \{(\Delta^k y_i \varepsilon_i)^2\} = \sigma^2 \sum_{j=0}^{k} C^2(k,j) = \sigma^2 \frac{(2k)!}{(k!)^2}$$

(The last step is to be proved as an exercise.)

In conclusion we see that the roundoff noise in the kth difference is approximately

$$\frac{(2k)!}{(k!)^2}$$

times as great as the noise in the original table, and hence the difference table, because it tends to emphasize the errors, should provide a means of forming an estimate of the roundoff error in a function.

k	$(2k)!/(k!)^2$
1	2
2	6
3	20
4	70
5	252
6	924
7	3,432
8	12,870

Our next difficulty is a standard one in statistics; we have a single table and have developed a theory for an ensemble of tables. The standard solution is to appeal to "the ergodic hypothesis" that **the average over the ensemble is the same as the average over the table**, of which we have a short, finite sample. All one can say in defense of this assumption is that it is both plausible and necessary.

The ergodic hypothesis is very commonly used by people without realizing it. For example, from an insurance mortality table giving the probabilities of 100,000 people's dying next year, one tries to deduce the pattern over a single life—a typical ergodic hypothesis application.

PROBLEMS 6.13

1 Expand both sides of

$$(1 + t)^{a+b} = (1 + t)^a (1 + t)^b$$

and equate like powers of t to get

$$C(a + b, r) = \sum_{s=0}^{a} C(a,s) C(b, r - s)$$

2 In the previous problem, set $a = b = r$ to get

$$C(2r,r) = \sum_{s=0}^{a} C^2(r,s)$$

3 In Prob. 1, let $a = b = n$, $r = n + 1$ to get

$$C(2n,\ n + 1) = \sum_{s=0}^{n} C(n,s)\, C(n,\ s - 1)$$

6.14 Correlation in the *k*th differences

We now have the pressing problem of deciding which difference column to use. To answer this, we need to develop another fact about kth differences that we actually use, namely, that adjacent values are highly correlated in a negative way. By correlation we refer to the obvious fact that the $\Delta^k y_i \varepsilon_i$ and $\Delta^k y_{i+1} \varepsilon_{i+1}$ use many of the same values of $y_j \varepsilon_j$.

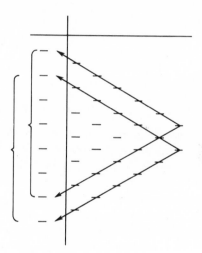

We therefore examine the quantity

$$\sum_{r=0}^{k} \frac{(-1)^r C(k,r)}{\sqrt{C(2k,k)}} \frac{(-1)^{r+1} C(k,\ r + 1)}{\sqrt{C(2k,k)}} = - \sum_{k=0}^{k} \frac{C(k,r)\, C(k,\ r + 1)}{C(2k,k)}$$

$$= - \frac{C(2k,\ k + 1)}{C(2k,k)} = - \frac{k}{k + 1}$$

This shows that if one term in a kth difference column is, say, $+$, then probably the next is $-$; our theory of roundoff noise estimation indicates that the higher k is, the more probable is the change in sign between successive entries.

k	$-k/k+1$	Probability †, %
0	$0 = 0$	50
1	$-\frac{1}{2} = -.50$	66.5
2	$-\frac{2}{3} = -.67$	74.2
3	$-\frac{3}{4} = -.75$	76.8
4	$-\frac{4}{5} = -.80$	79.4
5	$-\frac{5}{6} = -.83$	81.5
6	$-\frac{6}{7} = -.86$	82.8
7	$-\frac{7}{8} = -.875$	83.9
8	$-\frac{8}{9} = -.89$	84.7
9	$-\frac{9}{10} = -.90$	85.6
10	$-\frac{10}{11} = -.91$	86.3
20	$-\frac{20}{21} = -.953$	90.1

†The probabilities were found from suitable tables.

6.15 A roundoff estimation test

We now apply the various pieces of the theory. If the function were a polynomial, then the higher differences would all be zero (provided we went to a sufficiently high order of differences). We agreed that we usually compute at small enough steps so that in the absence of round-off errors the ideal differences tend to approach zero.

The second part of the theory is that if we have random data, we expect that a suitable column in the difference table will have alternating signs and that the differences will grow in size. We view our table as the sum of a table of the ideal numbers plus a table of the roundoff noise. The differences in the first table we hope go to zero;

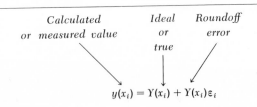

$$y(x_i) = Y(x_i) + Y(x_i)\varepsilon_i$$

We hope

1 $\Delta^k Y(x_i)$ *become small for increasing* k.
2 $\Delta^k Y(x_i)\varepsilon_i$ *become large for increasing* k.

those in the second tend to grow as we go to higher differences, and to alternate in sign in going down a particular column. We, therefore, in our desperation theory assume that when we see alternating signs in some kth difference column, we are seeing roundoff noise. Although counterexamples are easily constructed, they seem to be artificial, and in practice, this is a fairly safe guide.

How shall we test "alternating signs"? Experience shows that usually we operate around fourth and fifth differences where

$k = 4$	$k = 5$
$-\dfrac{k}{k+1} = \dfrac{-4}{5} - 0.80$	$\dfrac{-5}{6} - 0.83$
$p = 79.4\%$	$p = 81.5\%$

are the probabilities of a sign change between successive kth differences. Thus, if out of three possible changes in sign between four consecutive fourth or fifth differences we have at least two changes in sign, then we will say we are at "noise level." This is a practical compromise between opposing forces and represents an average criterion, not absolute safety. Having found k, we can now apply Sec. 6.13.

It is important to note that the structure of this roundoff theory does not depend on the particular calculation; we start with the answers and form a simple difference table from which we make our estimate. Thus, it is a theory that is independent of the particular calculation, though the results obtained depend on the particular numbers in the table.

PROBLEMS 6.15

1   Check the roundoff level in a standard five-place sin x table. Is it as expected?
2   Estimate roundoff level in the table

$$
\begin{array}{l}
21215 \\
21236 \\
21257 \\
21286 \\
21308 \\
21329
\end{array}
$$

INTEGRATION

7.1 Introduction

A common situation in engineering and science is that the answer to a problem is given by an integral that cannot be evaluated by the usual methods of analytic integration given in textbooks. There are two distinct kinds; the first kind is the indefinite integral

$$y(x) = \int_a^x f(t)\,dt$$

which requires a **table** of values to give the answer and which we shall take up in the next chapter as a special case of integrating an ordinary differential equation;

Indefinite integral

$$I(x,t) = \int_0^x \frac{e^{-tx^2}}{1+x^2}\,dx$$

where t is a parameter. For each t we need a table

x	$I(x,t)$
—	—
—	—
—	—
—	—

---

the second kind is the definite integral

$$y(b) = \int_a^b f(x)\,dx$$

which is a **single number** and is treated in this chapter.

Definite Integral

$$J(t) = \int_0^\infty \frac{e^{-tx^2}}{1+x^2}\, dx$$

For each t we need a single number.

There are two distinct approaches to the definite integral; the first is to use a single formula for numerically integrating from a to b; the second is to divide the range $a \leq x \leq b$ into a number of subintervals, apply a formula to each subinterval, and add together the results over each of the subintervals. The latter process generally uses the same formula in each subinterval and is called a **composite formula**.

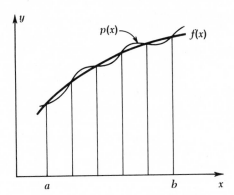

A single formula for the whole range

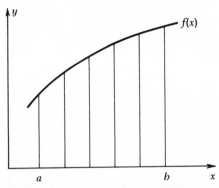

A formula (straight line) for each interval

The basic idea for finding an integral over a subinterval is to approximate the integrand $f(x)$ (or a part of the integrand) by a polynomial and then analytically integrate this, a process we called "analytic substitution" in Chap. 6.

$$\int_a^b f(x)\, dx \simeq \int_a^b p(x)\, dx$$

If we approximate the integrand with a polynomial of degree n, then obviously if the original function $f(x)$ is a polynomial of degree n or less, the formula will be exact for $f(x) = 1, x, \ldots, x^n$. Conversely, if the formula is exact for $f(x) = 1, x, \ldots, x^n$, then it will be true for any linear combination of them, namely, and polynomial of degree n.

These two not quite equivalent methods† both have their uses. The analytic substitution is a way of thinking about the problem; the "exact for $1, x, \ldots, x^n$" is a method for finding formulas easily.

Analytic substitution

Operate on $p(x)$ as if it were $f(x)$.

Exact

Make formula exactly true for

$$1, x, x^2, \ldots, x^n$$

There are so many different integration formulas that we shall only give a few of the more useful ones. By useful we require that they contain some method for estimating both the truncation and roundoff errors of the calculation or else have some other feature of considerable merit. These are severe, practical restrictions; however, the first one has led to the development of new formulas of some merit.

---

†Sometimes, it is not possible to find the approximating polynomial, e.g., given $y^{(-1)}, y^{(0)}, y^{(1)}, y''^{(-1)}, y''^{(0)}, y''^{(1)}$.

7.2 The trapezoid rule

If we approximate the function $f(x)$ in an interval $a \leq x \leq b$ by a straight line through the end points

$$y(x) = f(a) + \left[\frac{f(b) - f(a)}{b - a} \right] (x - a)$$

as we did in linear interpolation, and then integrate this approximation as if it were the original function, we get, after some algebra,

$$I = \int_a^b y(x)\, dx = \left[\frac{f(b) + f(a)}{2} \right] (b - a)$$

which is simply the classical formula for finding the area in a trapezoid.

Example   *The trapazoid rule by the exact matching method.*

We try

$$\int_a^b f(x) = w_1 f(a) + w_2 f(b)$$

and make it exact for both $f \equiv 1$ and $f \equiv x$.
For $f(x) \equiv 1$,

$$b - a = w_1 + w_2$$

For $f(x) = x$,

$$\frac{b^2 - a^2}{2} = w_1 a + w_2 b$$

Eliminate $w_2 b$:

$$\frac{b^2 - a^2}{2} - b(b - a) = w_1 a - w_1 b$$

$$\frac{b + a}{2} - b = -w_1$$

$$w_1 = \frac{b - a}{2}$$

Hence,

$$w_2 = b - a - w_1$$

$$= \frac{b - a}{2}$$

and the formula is

$$\int_a^b f(x)\, dx = \frac{b - a}{2}\,[f(a) + f(b)]$$

Usually the function $f(x)$ is such that a single straight line for the whole interval is not a good enough approximation. In such cases we can divide the interval $a \leq x \leq b$ into n subintervals,† each of length $h = (b - a)/n$, and apply the formula to each subinterval $(a, a + b)$,

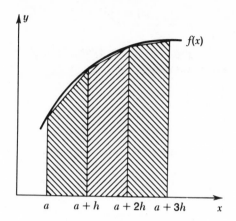

$(a + h, a + 2h), \ldots, (a + (n - 1) h, b)$ to get the composite formula

$$\int_a^b f(x) \, dx = h\left[\tfrac{1}{2}f(a) + f(a + h) + \cdots + \tfrac{1}{2}f(b)\right]$$

7.3 The truncation error in the trapezoid rule

We consider first the truncation error in the simple trapezoid formula.

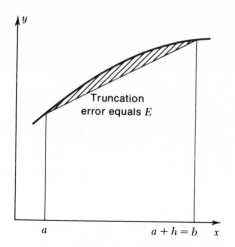

---

†We need not use the same spacing for each interval, but then the formula is not as elegant.

All we shall provide is an approximate theory (which, however, gives essentially the same result as the exact theory). We write the Taylor series for the integrand as

$$f(x) = f(a) + (x-a)f'(a) + \frac{(x-a)^2}{2!} f''(a) + \cdots$$

and substitute this in **both** sides of the simple trapezoid formula. The left-hand side is

$$\int_a^b \left[f(a) + (x-a)f'(a) + \frac{(x-a)^2}{2} f''(a) + \cdots \right] dx$$

which upon integration becomes

$$(b-a)f(a) + \frac{(b-a)^2}{2} f'(a) + \frac{(b-a)^3}{3!} f''(a) + \cdots$$

The right-hand side is, including the truncation error E,

$$\frac{h}{2}\left[f(a) + (b-a)f'(a) + \frac{(b-a)^2}{2} f''(a) + \cdots + f(a) \right] + E$$

Equating the two sides and recalling that $h = b - a$, we get, after cancellation of like terms,

$$\frac{(b-a)^3}{3!} f''(a) + \cdots = \frac{(b-a)^3}{4} f''(a) + \cdots + E$$

or

$$E = -\frac{(b-a)^3}{12} f''(a) + \cdots$$

The next term in the expansion has a $(b-a)^4$ factor, and we shall assume that **all** the terms beyond the first error term are small. Although it is not always true, for this formula it is true that the error E is exactly expressible in the form

$$E = -\frac{(b-a)^3}{12} f''(\theta) \qquad (a < \theta < b)$$

We next examine the composite formula. Here $(b-a)$ is h. Thus, the error term for n intervals is

$$-\frac{h^3}{12} f''(\theta_1) - \cdots - \frac{h^3}{12} f''(\theta_n) = \frac{-h^3}{12} \sum_{i=1}^{n} f''(\theta_i)$$

Now, by using the fact that†

$$\sum_{i=1}^{n} c_i f''(\theta_i) = \left(\sum_{i=1}^{n} c_i\right) f''(\theta)$$

provided all the $c_i \geq 0$, f'' is continuous, and θ is inside the range of the θ_i, the error term becomes

$$-\frac{h^3}{12} n f''(\theta) = -\frac{(b-a)h^2}{12} f''(\theta)$$

since nh is the new range $b - a$.

If we can bound the second derivative, then we have a bound on the truncation error. Unfortunately, such bounds either are not easily obtained or are useless. (See Probs. 4 and 5 below.)

PROBLEMS 7.3

1 Derive the compsite trapezoid rule for unequal spacing.
2 Derive the error term for Prob. 1 above.

Integrate by using the composite trapezoid rule, and estimate the truncation error in Probs. 3 to 6:

3 $\displaystyle\int_0^1 e^{-x^2}\,dx$ by using $h = \frac{1}{10}$

4 $\displaystyle\int_0^{\pi/2} \frac{\sin x}{x}\,dx$ by using $h = \dfrac{\pi}{18}$

5 $\displaystyle\int_0^1 x \ln x\,dx$ $h = \frac{1}{10}$

TRUE ANS. $\frac{1}{4}$

6 $\displaystyle\int_0^{\pi/2} \sin^2 x\,dx$ $h = \dfrac{\pi}{18}$

TRUE ANS. $\dfrac{\pi}{4}$

---

†This important result may be proved by induction with the use of

$$c_1 g(\theta_1) + c_2 g(\theta_2) = (c_1 + c_2)g(\theta)$$

where

$$\theta_1 < \theta < \theta_2$$

7.4   The midpoint formula

As an alternative to the trapezoid rule, suppose that instead of using the two end points of the interval we used the midpoint plus the slope at the midpoint to estimate the integral. Thus we seek a formula of the form

$$\int_a^b f(x)\,dx = w_1 f\left(\frac{a+b}{2}\right) + w_2 f'\left(\frac{a+b}{2}\right)$$

which represents the answer in terms of what we assume we know about $f(x)$.

First, we require that this formula be exact for $f(x) = 1$. This gives

$$b - a = w_1 + 0$$

Next we require that it be true for $f(x) \equiv x$ which leads to

$$\frac{b^2 - a^2}{2} = w_1 \frac{a+b}{2} + w_2$$

Solving these two equations, we get

$$w_1 = b - a$$

$$w_2 = 0$$

or $$\int_a^b f(x)dx = (b-a)f\left(\frac{a+b}{2}\right)$$

and we have our formula. Note that the $f'\left(\dfrac{a+b}{2}\right)$ term does not occur since $w_2 = 0$.

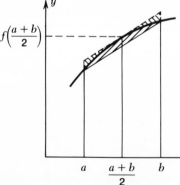

The line through the midpoint may be tilted *without* altering the area. When it has the slope of the tangent, the two small areas together are about half the large area.

To find the truncation error term, we again use the Taylor series

$$f(x) = f(a) + (x - a)f'(a) + \frac{(x - a)^2}{2} f''(a) + \cdots$$

We know in advance that the constant and linear terms will exactly cancel, and therefore, we need only substitute the quadratic term $[(x - a)^2/2] \, f''(a)$ into both sides to get the leading term of the error,

$$\frac{(b - a)^3}{6} f''(a) = \frac{b - a}{2} \left(\frac{b - a}{2}\right)^2 f''(a) + E$$

or

$$E = (b - a)^3 f''(a)[\ \tfrac{1}{6} - \tfrac{1}{8}\ ]$$
$$= \frac{(b - a)^3}{24} f''(a)$$

We note that the truncation error term for the midpoint formula is one-half as large as that for the trapezoid rule, has the opposite sign, and uses only one value of the function instead of two.

Archimedes (287 to 212 B.C.) knew that for $y = x^2$ the areas were exactly in the ratio of 1 to 2.

For the error term in the composite midpoint formula we have

$$\int_a^b f(x)\ dx = h\left[f\left(a + \frac{h}{2}\right) + f\left(a + \frac{3h}{2}\right) + \cdots + f\left(b - \frac{h}{2}\right)\right] + E$$

where, by the same reasoning as before,

$$E = \frac{(b - a)h^2}{24} f''(\theta)$$

PROBLEMS 7.4

1   Derive the midpoint formula by the analytic-substitution method.

2   Apply the midpoint formula to Prob. 3, Sec. 7.3.

3   Apply the midpoint formula to Prob. 5, Sec. 7.3.

7.5 Simpson's formula

Probably the most widely used formula for numerical integration is Simpson's formula

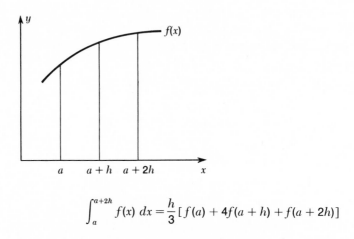

$$\int_a^{a+2h} f(x)\ dx = \frac{h}{3}[f(a) + 4f(a + h) + f(a + 2h)]$$

The corresponding composite formula is

$$\int_a^{a+2hn} f(x)\ dx = \frac{h}{3}[f(a) + 4f(a + h) + 2f(a + 2h)$$
$$+ 4f(a + 3h) + \cdots + f(a + 2nh)]$$

To derive these formulas, we can either approximate $f(x)$ by a quadratic through three points $x = a$, $a + h$, and $a + 2h$, (analytic-substitution method) and integrate this interpolating polynomial, or equivalently, make the formula exact for $f(x) = 1, x, x^2$ (the exact-matching method).   Using the latter, and setting $a = -h$ for convenience,

$$\int_{-h}^{h} f(x)\ dx = w_{-1}f(-h) + w_0 f(0) + w_1 f(1)$$

for

$$f \equiv 1: \qquad 2h = w_{-1} \qquad + w_0 \ + w_1$$
$$f \equiv x: \qquad 0 = -w_{-1}h \qquad\qquad + w_1 h$$
$$f \equiv x^2: \qquad \frac{2h^3}{3} = w_{-1}h^2 \qquad\qquad + w_1 h^2$$

From the middle equation $w_1 = w_{-1}$. Putting this into the last equation, we get

$$\frac{2h^3}{3} = 2w_1 h^2$$

$$w_1 = \frac{h}{3} = w_{-1}$$

Finally, the top equation gives

$$2h - \frac{2h}{3} = w_0 = \frac{4h}{3}$$

as it should.

When, to get the truncation error, we try $f(x) = x^3$, we get

$$0 = -w_{-1}h^3 + w_1 h^3 = 0$$

and we find that Simpson's formula is also exact for cubics!

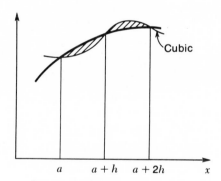

Areas exactly cancel for a cubic

To get the truncation error term, we use, therefore, the next term in the Taylor expansion, namely,

$$\frac{x^4}{4!} f^{iv}(\theta)$$

in both sides and obtain

$$\frac{2h^5}{5!} f^{iv} = \frac{h}{3}\left(\frac{h^4}{4!}f^{iv} + \frac{h^4}{4!}f^{iv}\right) + E$$

or

$$E = \frac{h^5}{4!} f^{iv}(\theta) \left(\frac{2}{5} - \frac{2}{3}\right)$$

$$= -\frac{h^5 f^{iv}(\theta)}{90}$$

For the composite Simpson formula, the error term is

$$-\frac{(b-a)h^4 f^{iv}(\theta)}{90}$$

This shows why Simpson's formula is so popular.

PROBLEMS 7.5

1 Derive the three-eighth's rule

$$\int_a^{a+3h} f(x)\, dx = \frac{3h}{8}\left[\, f(a) + 3f(a+h) + 3f(a+2h) + f(a+3h)\right]$$

2 Derive the error term, and compare with Simpson's.

3 Develop the composite three-eighth's rule.

Integrate Probs. 4 to 7 by using Simpson's formula:

4 $\displaystyle\int_0^1 e^{-x^2}\, dx \qquad h = \frac{1}{10}$

5 $\displaystyle\int_0^\pi \frac{1 - \cos x}{x}\, dx \qquad h = \frac{\pi}{9}$

6 $\displaystyle\int_0^1 \sqrt{1 - x^2}\, dx \qquad h = \frac{2}{10}$

7 $\displaystyle\int_0^2 \frac{e^{-t}}{1 + t^2}\, dt \qquad h = \frac{2}{10}$

7.6 Some practical details

If we use the composite trapezoid, midpoint, or Simpson's formula, and if we want an estimate of the error due to truncation, then we must estimate a derivative, in these cases either the second or fourth derivative. In practice this derivative is rarely available.

More often than most people suspect, in practical problems the higher derivatives will become infinite in the range, and thus, the error term will not provide a practical bound or else it will be extremely hard to find the finite bound. (See various problems for illustration.)

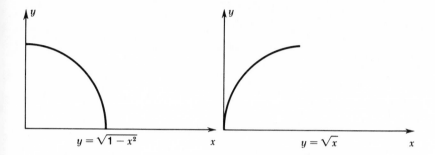

$$y = \sqrt{1 - x^2} \qquad\qquad y = \sqrt{x}$$

One solution to this dilemma is to carry out the formula for the spacing h and the spacing $h' = h/2$, and compare the answers. In the composite trapezoid rule the error should drop, roughly, by a factor 4, whereas the error in Simpson's composite formula should drop by a factor of 16. Thus, if at the two spacings h and $h/2$, the answer does not change significantly, then the change in the truncation error is too small to be observed and we can hope that we have an accurate answer.

Three stages h, $h' = h/2$, and $h'' = h/4$ will usually provide more safety. From the three trials we get

$$I = S + E = S' + E' = S'' + E''$$

where we expect, for example, for Simpson's composite formula,

$$E = 16E' = 256E''$$

Thus,

$$\frac{S' - S}{15} = E' \qquad \frac{S'' - S'}{15} = E'' = \frac{1}{16}E'$$

and if the numbers $S' - S$ and $S'' - S'$ are in the approximate ratio of 16 to 1, then this indicates that our error estimates are probably correct.

But there is always the trouble that we face both truncation **and** roundoff errors and the latter may mask the truncation error. For this reason we will in the next three sections develop a different class of formulas which tries to separate the two effects.

PROBLEMS 7.6

Estimate the error by using h, $h/2$, and $h/4$ for the following integrals:

1 $\displaystyle\int_0^1 e^{-x^2}\,dx \qquad h = \tfrac{1}{4}$

2 $\displaystyle\int_0^1 \sqrt{1-x^2}\,dx \qquad h = \tfrac{1}{2}$

3 $\displaystyle\int_0^2 \frac{e^{-t}}{1+t^2}\,dt \qquad h = \tfrac{1}{2}$

4 $\displaystyle\int_0^1 x \ln x\,dx \qquad h = \tfrac{1}{4}$

7.7 The Gregory type of formula

A convenient formula for numerical integration which provides separate roundoff and truncation error estimates is the Gregory formula

$$\int_{-n}^{n} f(x)\,dx = \tfrac{1}{2}f(-n) + f(-n+1) + f(-n+2) + \cdots + \tfrac{1}{2}f(n) + E_T$$

where

$$E_T = -\tfrac{1}{12}\left[\Delta f(n-1) - \Delta f(-n)\right] - \tfrac{1}{24}\left[\Delta^2 f(n-2) + \Delta^2 f(-n)\right] + \cdots$$

A DIFFERENCE TABLE

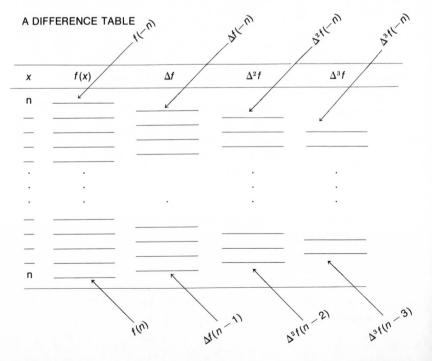

This formula is effectively the trapezoid rule plus end corrections of as high an order as desired. Evidently, if we include, say, the kth differences, then the formula will be exact for polynominals of degree k or less.

The size of the first terms neglected will provide an estimate of the truncation error, whereas the differences in the table will provide an estimate of the roundoff error (if we use the results of Chap. 6).

The range $-n$ to n is easily changed to any other range and spacing by a simple transformation. The fact that we have used a symmetrical interval is not essential, and the formula also holds for an odd number of intervals.

The Gregory formula is a special case of a recently discovered class of such formulas,† and we shall derive the Gregory formula as a special case. The general class was found by deliberately searching for formulas having covenient separate estimates of both roundoff and truncation error.

7.8 The general formula

We shall examine a class of formulas of the form

$$\int_{-n}^{n} f(x)\,dx = af(-n) + bf(-n + 1) + cf(-n + 2) + bf(-n + 3)$$
$$+ cf(-n + 4) + \cdots + bf(n - 1) + af(n)$$
$$+ \sum_{s=1}^{\infty} q_s \left[\Delta^s f(n - s) + (-1)^s \Delta^s f(-n) \right]$$

where the q_s do not depend on n. We note that for $a = \frac{1}{2}$ with $b = c = 1$, this is Gregory's formula.

If this formula is true for $f(x) \equiv 1$, then necessarily

$$2n = 2a + nb + (n - 1)c$$

If the weights a, b, and c are independent of n, which is one of the nice properties of the Gregory formula, then

$$(b + c - 2)n + (2a - c) \equiv 0 \text{ implies}$$

$$b + c - 2 = 0$$
$$2a - c = 0$$

†Hamming, R. W., and Pinkham, R. S., A Class of Integration Formulas, *JACM*, vol. 13, no. 3 (July, 1966), pp. 430–438.

or

$$c = 2a$$
$$b = 2 - 2a$$

Note that for each a we have a different formula. We therefore have a one-parameter family of formulas. As noted, $a = \frac{1}{2}$ gives the composite trapezoid rule plus end correction terms. For $a = \frac{1}{3}$, we get Simpson's composite formula plus end correction terms. We must still determine the q_s as functions of the parameter a.

If

$$a = \tfrac{1}{3}$$
$$b = \tfrac{2}{3}$$
$$c = \tfrac{4}{3}$$

then

$$(a, b, c, b, c, \ldots, b, a)$$

is

$$\tfrac{1}{3}(1, 4, 2, 4, 2, \ldots, 4, 1)$$

which is Simpson's composite formula.

If the formula is to be exact for 1, x, ..., x^k, ..., then

$$\int_{-n}^{n} x^k \, dx = a(-n)^k + b(-n+1)^k + \cdots + a(n)^k + \sum_{s=1}^{\infty} q_s [\Delta^s(n-s)^k + (-1)^s \Delta^s(-n)^k] \quad (k = 0, 1, \ldots)$$

Appendix A derives the formulas for the coefficients q_s.

7.9 Results

In Appendix A to this chapter the following table of the coefficients q_s of Δ^s is derived.

The first column gives the coefficients for the Gregory formula we started with.

In the second column, we have the coefficients for "Simpson's formula plus end corrections." Note that the coefficients are all smaller in size than for the trapezoid rule of Gregory's formula. In this sense we can say that this formula is "better than" the Gregory

s	Gregory, $a = \frac{1}{2}$	Simpson, $a = \frac{1}{3}$	$a = 0$
1	$-\frac{1}{12}$	0	$\frac{2}{12}$
2	$-\frac{1}{24}$	0	$\frac{2}{24}$
3	$-\frac{19}{720}$	$-\frac{4}{720}$	$\frac{26}{720}$
4	$-\frac{9}{480}$	$-\frac{4}{480}$	$\frac{6}{480}$
5	$-\frac{863}{60,480}$	$-\frac{548}{60,480}$	$\frac{82}{60,480}$
6	$-\frac{275}{24,192}$	$-\frac{212}{24,192}$	$-\frac{86}{24,192}$
7	$-\frac{33,953}{3,628,800}$	$-\frac{29,228}{3,628,800}$	$-\frac{19,778}{3,628,800}$

formula; however, there always exist special integrands for which a particular integration formula is exactly correct and for which a "better formula" will do worse.

In the third column, $a = 0$, we have the very interesting case in which half the coefficients are zero,

---

Remember,

$$a = 0$$

implies

$$\begin{cases} c = 2a = 0 \\ b = 2 - 2a = 2 \end{cases}$$

---

and many of these function values never need to be computed! For $a = 0$ the first few q_s are positive and larger than those of the Simpson formula, but if we include the fourth differences as end correction terms, then the rest of the coefficients are smaller than either of the other two! Thus $a = 0$ seems especially attractive.

We need to compute a number of the end values at the regular spacing because we need the values for the differences for the end corrections and for the roundoff estimates,

x	$f(x)$	$\Delta f(x)$	$\Delta^2 f$	$\Delta^3 f$
a	—			
		—		
b	—		—	
		—		—
c	—		—	
		—		
b	—			
c	o†			
b	—			
c	o†			
b	—			
c	o†			

†Can be omitted if only Δ^3 is used.

but over the middle range we do not need half the integrands since they enter into the formula with zero coefficients.

Experience shows that typically we can use the fourth (or some-times the fifth) differences to estimate the roundoff error, and if we are not down to roundoff noise, then we can look at the first Δ^k's we dropped and ask if the truncation error is small enough. If the trunca-tion error is not small enough, then we decrease the interval by a fac-tor of $\frac{1}{3}$ for the $a = 0$ (thus, we can use the old points) and of $\frac{1}{2}$ for the $a = \frac{1}{3}$.

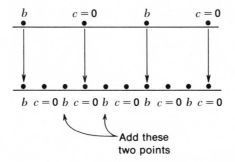

If we keep the fourth differences (and roundoff noise is not causing trouble), then the truncation error should drop by a factor of about $3^k = 243$ for $a = 0$, and by a factor of $2^k = 32$ for $a = \frac{1}{3}$.

If we meet roundoff errors before we reduce the truncation to the size we wish, then we can try fewer differences at that spacing and hope for the best, but generally we are in real trouble.

PROBLEMS 7.9

Apply the appropriate integration formula using at least five-decimal arithmetic and a suitable number of differences:

1 $\displaystyle\int_0^1 e^{-x^2}\,dx$ $h = \frac{1}{10}$ $a = 0$

2 $\displaystyle\int_0^\pi \sin^2 x\,dx$ $h = \dfrac{\pi}{8}$ $a = 0$

3 $\displaystyle\int_0^\pi \sin^2 x\,dx$ $h = \dfrac{\pi}{8}$ $a = \frac{1}{3}$

4 $\displaystyle\int_0^2 \frac{e^{-x}}{1 + x^2}\,dx$ $h = \frac{2}{10}$ $a = 0,\ \frac{1}{3},\ \frac{1}{2}$

7.10 Gauss quadrature

Up to now we have often used $n + 1$ equally spaced sample points and fitted a polynomial of degree n, using the $n + 1$ sample points to determine the coefficients. As a result, the formula was exact for polynomials of degree n, though as in Simpson's formula it sometimes happens that the formula is exact for polynomials of one degree higher.

The idea behind Gauss type of formulas is that by also using the **positions** of the sample point as parameters, we can get a formula by using n samples which is exact for polynomials of degree up to and including $2n - 1$. For many purposes this is a significant gain. The theory, however, is so complex that we will give only one example of a Gauss quadrature formula.

Consider the formula

$$\int_{-1}^1 f(x)\,dx = w_1 f(x_1) + w_2 f(x_2)$$

where the w_i and the x_i are **all** parameters. Thus, we can try to make the formula exact for

$$f = 1,\ x,\ x^2,\ x^3$$

$$f = 1: \qquad 2 = w_1 + w_2$$

$$f = x: \qquad 0 = w_1 x_1 + w_2 x_2$$

$$f = x^2: \qquad \tfrac{2}{3} = w_1 x_1^2 + w_2 x_2^2$$

$$f = x^3: \qquad 0 = w_1 x_1^3 + w_2 x_2^3$$

We need to solve these four nonlinear equations for our unknowns w_1, w_2, x_1, and x_2. It is easy to see that if we pick

$$w_1 = w_2$$

$$x_1 = -x_2$$

then both the second and fourth equations are satisfied. The other two equations become

$$2 = 2w_1 \qquad \text{or} \qquad w_1 = 1$$

and

$$\tfrac{2}{3} = 2w_1 x_1^{\,2} \qquad \text{or} \qquad x_1^{\,2} = \tfrac{1}{3}$$

Hence,

$$x_1 = \frac{\pm 1}{\sqrt{3}} = \pm 0.577 \cdots$$

Our formula is, therefore,

$$\int_{-1}^{1} f(x)\ dx = f\left(\frac{-1}{\sqrt{3}}\right) + f\left(\frac{1}{\sqrt{3}}\right)$$

and it is exact for cubics in spite of the fact that it uses only two sample points!

There are three classic forms of Gauss quadrature,

$$\int_{-1}^{1} f(x)\ dx \qquad \int_{0}^{\infty} e^{-x} f(x)\ dx \qquad \int_{-\infty}^{\infty} e^{-x^2} f(x)\ dx$$

in each of which the $f(x)$ is approximated by a polynomial. Extensive tables of the sample points x_i and the corresponding weights w_i are available.†

---

†Handbook of Mathematical Functions, *NBS Appl. Math.*, Ser. 55, 1964, pp. 916–924.

PROBLEMS   7.10

Find the Gauss quadrature formula for the following:

1 $\displaystyle\int_{-1}^{1} f(x) = w_1 f(x_1) + w_2 f(x_2) + w_3 f(x_3)$

NOTE: $x_1 = x_3 \quad w_1 = w_3$
$\ x_2 = 0$

2 $\displaystyle\int_{0}^{\infty} e^{-x} f(x)\ dx = w_1 f(x_1) + w_2 f(x_2)$

3 $\displaystyle\int_{-\infty}^{\infty} e^{-x^2} f(x)\ dx = w_1 f(x_1) + w_2 f(x_2) + w_3 f(x_3)$

APPENDIX

7A.1   The basic derivation

We are going to solve the system of equations

$$\int_{-n}^{n} x^k\ dx = a(-n)^k + b(-n+1)^k + \cdots + a(n)^k + \sum_{s=1}^{\infty} q_s [\Delta^s (n-s)^k +$$
$$(-1)^s\ \Delta^s (-n)^k] \qquad (k = 0, 1, \ldots)$$

by the "generating function method." In this method we multiply the kth equation by $t^k/k!$ and sum all the equations. Using the fact that

$$\sum_{k=0}^{\infty} \frac{x^k t^k}{k!} = e^{xt}$$

we get

$$\int_{-n}^{n} e^{xt}\ dx = ae^{-nt} + be^{(-n+1)t} + ce^{(-n+2)t} + \cdots + ae^{nt}$$
$$+ \sum_{s=n}^{\infty} q_s [\Delta^s e^{(n-s)t} + (-1)^s\ \Delta^s e^{-nt}]$$

But

The symbol $\Delta_x e^{xt}$ means that x is the variable that is used in the differencing.

$$\Delta_x e^{xt} = e^{(x+1)t} - e^{xt} = e^{xt}(e^t - 1)$$

hence,

$$\Delta_x{}^s e^{xt} = e^{xt}(e^t - 1)^s$$

We have, therefore,

$$\int_{-n}^{n} e^{xt}\,dx = a(e^{-nt} + e^{nt}) + b \sum_{k=0}^{n-1} e^{(-n+1+2k)t} + c \sum_{k=1}^{n-1} e^{(-n+2k)t}$$
$$+ \sum_{s=1}^{\infty} q_s[e^{nt}(1 - e^{-t})^s + e^{-nt}(1 - e^t)^s]$$

Doing the geometric sums and the integration, we get

$$\frac{e^{nt} - e^{-nt}}{t} = a(e^{nt} + e^{-nt}) + b\,\frac{e^{nt} - e^{-nt}}{e^t - e^{-t}} + c\,\frac{e^{(n-1)t} - e^{-(n-1)t}}{e^t - e^{-t}}$$
$$+ \sum_{s=1}^{\infty} q_s[e^{nt}(1 - e^{-t})^s + e^{-nt}(1 - e^t)^s]$$

Notice that we can divide this **identity** in t into two parts. The first part is

$$\frac{e^{nt}}{t} = ae^{nt} + \frac{be^{nt}}{e^t - e^{-t}} + \frac{ce^{(n-1)t}}{e^t - e^{-t}} + \sum_{s=1}^{\infty} q_s e^{nt}(1 - e^{-t})^s$$

and the second part is exactly the same except that t is replaced by $-t$. Thus this single identity in t is all we need use to determine the constants q_s.

This expression is an expansion in powers of $1 - e^{-t}$, so we set

$$v = 1 - e^{-t}$$

which is the same as

$$t = -\ln(1 - v)$$

and factor out the e^{nt}, leaving (after some algebra)

$$\frac{-1}{\ln(1 - v)} = a + \frac{b(1 - v)}{v(2 - v)} + \frac{c(1 - v)^2}{v(2 - v)} + \sum_{s=1}^{\infty} q_s v^s$$

Rearranging this, we have

$$\sum_{s=1}^{\infty} q_s v^s = -\left[\frac{1}{\ln(1 - v)} + a + \frac{b(1 - v)}{v(2 - v)} + \frac{c(1 - v)^2}{v(2 - v)}\right]$$

and we see that q_s is simply the coefficient of v^s in the power series expansion of the right-hand side.

7A.2   The Gregory formula

Before doing the general formula, we turn to the special Gregory formula. In this case $a = \frac{1}{2}$ and $b = c = 1$. Our formula for the q_s is therefore

$$\sum_{s=1}^{\infty} q_s v^s = -\left[\frac{1}{\ln(1-v)} + \frac{1}{2} + \frac{1-v}{v}\right]$$

$$= -\left[\frac{1}{\ln(1-v)} - \frac{1}{2} + \frac{1}{v}\right]$$

The troublesome part is finding the power series for the reciprocal of the log term. To get this, we start by noting

$$\ln(1-v) = -\int_0^v \frac{dv}{1-v} = -\int_0^v \sum_{k=0}^{\widetilde{~}} v^k \, dv$$

$$= -v - \frac{v^2}{2} - \frac{v^3}{3} - \cdots$$

$$= -v\left(1 + \frac{v}{2} + \frac{v^2}{3} + \cdots\right)$$

When we divide this into 1 and combine it with the two other terms, we get

$$\sum_{s=1}^{\infty} q_s v^s = -\frac{v}{12} - \frac{v^2}{24} - \frac{19}{720} v^3 - \cdots$$

More values of the coefficients are given in the table in the text under Gregory, $a = \frac{1}{2}$. These coefficients are the q's, the coefficients of the differences in the integration formula.

7A.3   Other special cases

If we write the coefficients of Gregory's formula as g_s, we have from the definition

$$-\frac{1}{\ln(1-v)} = \frac{1}{v} - \frac{1}{2} + \sum_{s=1}^{\infty} g_s v^s$$

This can be used in the general formula at the end of Sec. A.3 to eliminate the term

$$-\frac{1}{\ln(1-v)}$$

to get the convenient formula

$$\sum_{s=1}^{\infty} q_s v^s = \frac{1}{v} - \frac{1}{2} + \sum g_s v^s - a - \frac{b(1-v)}{v(2-v)} - \frac{c(1-v)^2}{v(2-v)}$$

Now, using $c = 2a$, $b = 2 - 2a$, and some routine algebra

$$\sum_{s=1}^{\infty} q_s v^s = \sum_{s=1}^{\infty} g_s v^s + \left(\frac{1}{2} - a\right)\frac{v}{(2-u)}$$

Dividing out this last term,

$$\frac{v}{2-v} = \frac{v/2}{1-u/2} = \sum_{s=1}^{\infty}\left(\frac{v}{2}\right)^s$$

we get finally (upon equating like powers of v^s)

$$q_s = g_s + \left(\frac{1}{2} - a\right)\frac{1}{2^s}$$

from which the table entries follow easily.

ORDINARY DIFFERENTIAL EQUATIONS

8

8.1 Meaning of a solution

Differential equations are of frequent occurrence in engineering, and relatively few of those that appear can be solved in closed form by the standard mathematical tricks. We therefore need to consider the numerical solution of ordinary differential equations.

Given a single first-order ordinary differential equation

$$y' = \frac{dy}{dx} = f(x,y)$$

what do we mean by a solution?

The indefinite integral

$$y(x) = \int_a^x f(t)\ dt$$

is the special differential equation

$$y'(x) = f(x)$$

where $f(x,y)$ does not depend on y.

Loosely speaking, we mean a curve

$$y = y(x)$$

such that if we calculate the value of the derivative of $y(x)$ at a point (x,y), then this value will be the same as that given by the differential equation [also evaluated at the same point (x,y)].

This suggests a crude method for numerically solving a given differential equation. At each point (x,y) of a rectangular mesh of points, we calculate the slope from the given differential equation and then draw through each point a short line segment which has the calculated slope.

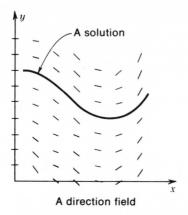

A direction field

Such a picture is called a **direction field**. Our problem now is to draw a smooth curve through the direction field such that the curve is always tangent to the appropriate lines of the direction field. If we succeed, then we must have a solution of the given differential equation.

We see immediately that through almost any point in the area of the direction field, we can draw a curve. Thus, we are given normally both the equation and a point through which the curve is to pass.

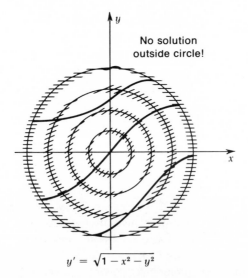

$$y' = \sqrt{1 - x^2 - y^2}$$

This point is usually called the **initial condition**.

The curves along which the slope has a constant value k are called **isoclines** and are defined by

$$k = f(x,y)$$

Often this provides an easy way to draw a direction field. Note that the maxima and minima of the solution must lie on the zero isocline.

The direction-field approach just described, though simple, is very useful and practical. On occasions a quick sketch of a direction field can settle an important point under discussion.

PROBLEM 8.1

1 Show that the inflection points of solutions lie on the curve $y'' = 0$.

8.2 Improved direction-field method

If we want only a single curve (solution), then it becomes immediately clear that we need draw the direction field only in the region where the solution is going to go and we can ignore all the rest of the (x,y) plane.

Thus, we start at the **initial point** (x_0,y_0), calculate the slope $y'_0 = f(x_0,y_0)$, and go a "short distance" in this direction to a second point. We now regard this second point as a new initial point and repeat the process again.

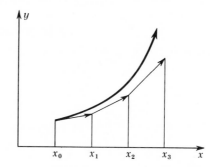

After enough small steps we shall have the solution on the finite interval we were interested in. If we want to know how the solution behaves at infinity, we shall probably be satisfied (after a moderate distance) by the general behavior of the direction field far out, or else we shall have to make a change of variables to transform infinity to some finite point.

It should be also clear that the smaller the steps we take, the less truncation error there will be but the more effort we must expend and the greater the possible roundoff error will be. It is obvious that we need not make the drawing; we can merely tabulate on a sheet of paper the coordinates of the points. Then, given a point (x_n, y_n) on the solution curve, the process is

1   Calculate $y'_n = f(x_n, y_n)$, which is the slope of the curve at the point.

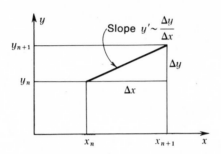

2   Calculate the next point

$$y_{n+1} = y_n + hy'_n$$

$$x_{n+1} = x_n + h$$

And repeat steps 1 and 2 until we reach the end of the interval. We have chosen a uniform step size $\Delta x_n = h$ for convenience.

Method applied to

$$y' = \sqrt{1 - x^2 - y^2}$$
$$y(0) = 0 \qquad h = \tfrac{1}{10}$$

With two decimal places except in the values of y

x	y	$x^2 + y^2$	y'
0	0	$0 + 0 = 0$	$\sqrt{1} = 1$
0.1	0.100	$0.01 + 0.01 = 0.02$	$\sqrt{0.98} = 0.99$
0.2	0.199	$0.04 + 0.04 = 0.08$	$\sqrt{0.92} = 0.96$
0.3	0.295	$0.09 + 0.09 = 0.18$	$\sqrt{0.82} = 0.91$
0.4	0.386	$0.16 + 0.15 = 0.31$	$\sqrt{0.69} = 0.83$
0.5	0.469	$0.25 + 0.22 = 0.47$	$\sqrt{0.53} = 0.73$
0.6	0.572	$0.36 + 0.29 = 0.65$	$\sqrt{0.35} = 0.59$
0.7	0.601	$0.49 + 0.36 = 0.85$	$\sqrt{0.15} = 0.39$
0.8	0.640	$0.64 + 0.41 = 1.05$	*imaginary*
0.9			
1.0			

TABLE

x	y	y'	hy'
–	–	–	–
–	–	–	–
–	–	–	–
–	–	– ·	–

PROBLEMS 8.2

1 Integrate

$$y' = y^2 + x^2 \qquad (0 \leq x \leq 1)$$

starting at $x = 0$, $y = 0$ and using $h = 0.1$.

2 Integrate

$$y' = e^{-y} - x^2 \qquad (0 \leq x \leq 2)$$

through (0,0), using $h = 0.2$.

3 Integrate

$$y' = \sin y - x \qquad \left(0 \leq x \leq \frac{\pi}{2}\right)$$

$y(0) = 0$, using $h = \pi/12$.

8.3 Modified Euler method

The trouble with the preceding method is that we tend to make systematic truncation errors because we are generally using the derivative of the curve that applied somewhat before where we are now. The following method, called the "modified Euler method" gives a better (more economical and more practical) method of calculation.

We suppose for the moment that we have a pair of points (x_{n-1}, y_{n-1}) and (x_n, y_n) and ask, "How shall we find the next point (x_{n+1}, y_{n+1})?"

We will **predict** the next value by applying the midpoint formula to

$$\int_{x_{n-1}}^{x_{n+1}} y'(x)\, dx,$$

which gives

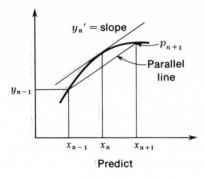

Predict

$$p_{n+1} = y_{n-1} + 2hy_n'$$

where p_{n+1} is the predicted value of y_{n+1}. Clearly, we are using the tangent line at the middle of the double interval as a guide for how to move across it.

Using this predicted value, we compute the slope at the predicted point

$$p'_{n+1} = f(x_{n+1}, p_{n+1})$$

and apply the trapezoid rule to $\int_{x_n}^{x_{n+1}} y'(x)\, dx$ which gives

$$y_{n+1} = y_n + \frac{h}{2}\,(p'_{n+1} + y'_n)$$

as the **corrected** value y_{n+1}. Here we are using the average of the slopes at the two ends of the interval of integration as the average slope in the interval.

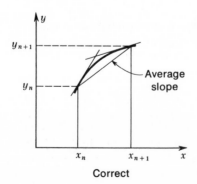

Correct

We are now ready for the next step. In both the predictor and the corrector we have avoided some of the systematic error we discussed for the simple method.

The method has four steps

$$p_{n+1} = y_{n-1} + 2hy_n{}'$$

$$p'_{n+1} = f(x_{n+1}, p_{n+1})$$

$$y_{n+1} = y_n + \frac{h}{2}(y_n{}' + p'_{n+1})$$

$$y'_{n+1} = f(x_{n+1}, y_{n+1})$$

8.4 Starting the method

We assumed that we had two starting values (x_{n-1}, y_{n-1}) and (x_n, y_n), but when the problem comes to us we are given only one starting point. We shall give two methods of starting, one suitable for hand calculation and one for machine calculation.

The hand-calculation method of starting is based on the Taylor expansion

$$y(x + h) = y(x) + hy'(x) + \frac{h^2}{2!} y''(x) + \cdots$$

The derivatives are easily found from the differential equation by differentiating

$$y' = f(x,y)$$

$$y'' = \frac{\partial f}{\partial x} + \frac{\partial f}{\partial y} y'$$

$$y''' = \text{etc.}$$

and then evaluating the derivatives. The number of terms to take depends on the step size h and the accuracy desired.

Example

$$y' = y^2 + x^2$$

through the point (0,0)

$$y'' = 2yy' + 2x$$
$$y''' = 2y'^2 + 2yy'' + 2$$
$$y^{iv} = 6y'y'' + 2yy'''$$

at (0,0)

$$y = 0$$
$$y' = 0$$
$$y'' = 0$$
$$y''' = 2$$
$$y^{iv} = 0$$

Therefore,

$$y(h) = 0 + h0 + \frac{h^2}{2}0 + \frac{h^3}{3!}2 + \cdots$$

$$= \frac{h^3}{3} + \cdots$$

The second method which is more suitable for machine calculation is based on the repeated use of the corrector. Given (x_0, y_0), we first estimate an earlier point

$$(x_{-1}, y_{-1})$$

by the linear approximation

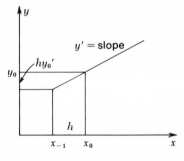

$$x_{-1} = x_0 - h$$

$$y_{-1} = y_0 - hy_0'$$

$$y_{-1}' = f(x_{-1}, y_{-1})$$

We then use the corrector backward

$$y_{-1} = y_0 - \frac{h}{2}(y_0' + y_{-1}')$$
$$y_{-1}' = f(x_{-1}, y_{-1})$$

and repeat this backward corrector enough times so that the value of y_{-1} settles down. If it does not settle down within a few trials, then h should be decreased. The value y_{-1} is used only once, in the first prediction.

Example
$$y' = y^2 + x^2 \qquad point \ (0,0)$$
$$x_{-1} = 0 - h = -h$$
$$y_{-1} = 0 - h0 = 0$$
$$y_{-1}' = h^2$$

Use corrector

$$y_{-1} = 0 - \frac{h}{2}(0 + h^2)$$

$$= -\frac{h^3}{2}$$

$$y_{-1}' = h^2 + \frac{h^6}{4} \sim h^2$$

NOTE:
1   *We got $y_{-1} = -h^3/2$ and not, as we should have, $-h^3/3$.*
2   *Iteration on the corrector will not change either y_{-1} or y'_{-1} in this case.*

8.5   Error estimates

The predictor, as noted, is the midpoint formula (Sec. 7.3) with the use of a double interval, so that the error is

$$E = \frac{(2h)^3 y'''(\theta)}{24} = \frac{h^3 y'''(\theta)}{3}$$

On the other hand, the error in the corrector (Sec. 7.2) is

$$E = \frac{-h^3}{12} y'''(\theta_2)$$

We are integrating y', not y. So the derivatives are 1 higher than in Chap. 7.

Thus, the predictor value minus the corrector value is

$$p - c = \left[\text{true} - \frac{h^3}{3} y'''(\theta_1) \right] - \left[\text{true} + \frac{h^3}{12} y'''(\theta_2) \right]$$

Assuming that the third derivative does not change much in the double interval (and if it did, we should have to decrease h until it did not), then $y'''(\theta_1) \simeq y'''(\theta_2)$ and we have

$$p - c = -\tfrac{5}{12} h^3 y'''(\theta)$$

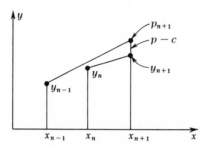

Thus, in a computation we naturally monitor the quantity $p_{n+1} - c_{n+1}$ where what we wrote as y_{n+1} is now labeled c_{n+1}, the corrected value. From these values we can deduce that the actual truncation errors are probably about $-\tfrac{4}{5}(p_{n+1} - c_{n+1})$ in the predictor and $\tfrac{1}{5}(p_{n+1} - c_{n+1})$ in the corrector.

Summary

Predict

$$p_{n+1} = y_{n-1} + 2hy'_n$$

Evaluate

$$p'_{n+1} = f(x_{n+1}, p_{n+1})$$

Correct

$$y_{n+1} = y_n + \frac{h}{2}(y'_n + p'_{n+1})$$

Evaluate

$$y'_{n+1} = f(x_{n+1}, y_{n+1})$$

We have played a bit fast and loose in our arguments. For example, the error in the trapezoid rule is based on the assumption that we have the correct y'_{n+1} value. If $p_{n+1} - c_{n+1}$ were large, then we should have cause to doubt that this is true and we should be tempted to iterate the corrector several times. As a practical matter it is probably better to shorten the interval than it is to iterate the corrector.

The following is a slight elaboration of the basic method, which somewhat increases the accuracy and is based on the observation that the entries in the $p - c$ column change slowly (if they do not, then we shall have to shorten the interval). Since we recognize that $p_{n+1} - c_{n+1}$ gives us an estimate of the error, it is tempting to use this information to make small corrections as we go along. Thus, when we **predict**

$$p_{n+1} = y_{n-1} + 2hy_n$$

we suspect that the **modified** value

$$m_{n+1} = p_{n+1} - \tfrac{4}{5}(p_n - c_n)$$

would be more accurate than the predicted value. We then compute

$$m'_{n+1} = f(x_{n+1}, m_{n+1})$$

and **correct**, using

$$c_{n+1} = y_n + \frac{h}{2}\left(m'_{n+1} + y'_n\right)$$

We take as the final value

$$y_{n+1} = c_{n+1} + \tfrac{1}{5}(p_{n+1} - c_{n+1})$$

and compute

$$y'_{n+1} = f(x_{n+1}, y_{n+1})$$

In doing the computation by hand, we should use a table of the form

TABLE

x_n	y_n	y'_n	p_n	m_n	m'_n	c_n	$p_n - c_n$
—	—	—	—	—	—	—	—
—	—	—	—	—	—	—	—

8.6 An example using Euler's method

As an example of the preceding, consider the differential equation

$$y' = y^2 + 1$$
$$y(0) = 0$$

To get started by the machine-calculation method with $h = \frac{2}{10}$ and $y'(0) = 1$, we get from the linear extrapolation

$$y(-0.2) = 0 - 0.2(1) = -0.2$$

$$y'(-0.2) = 1.04$$

Then the backward corrector

$$y(-0.2) = 0 - 0.1(1 + 1.04)$$
$$= -0.204$$
$$y'(-0.2) = 1.042$$

and we can see that further repetition will produce negligible improvement.

Let us do the first step in some detail. Predict

$$p(0.2) = -0.204 + 0.4(1) = 0.196$$

$$p'(0.2) = 1 + (0.196)^2 = 1.038$$

$$c(0.2) = 0 + 0.1(1 + 1.038) = 0.204$$

The initial step needs to be especially carefully done; otherwise we start wrong and continue to go wrong!

Since $p - c = 0.196 - 0.204 = -0.008$ is rather large, we repeat the first corrector step to be safe.

$$c'(0.2) = 1 + (0.204)^2 = 1.042$$

$$c(0.2) = 0 + 0.1(1 + 1.042) = 0.204$$

With no change, we feel somewhat safe and go on to the next step.

$$p(0.4) = 0 + 0.4(1.042) = 0.417$$

$$p'(0.4) = 1 + (0.417)^2 = 1.174$$

$$c(0.4) = 0.204 + 0.1(1.042 + 1.174) = 0.426$$

Now $p - c = -0.009$, so that the error is around 0.002 on the $c(0.4)$ value, and if this error is tolerable for each step (remember it accumulates along the solution), then we can continue; otherwise we should restart, using a smaller h, say, $h = 0.10$ or $h = 0.15$.

PROBLEMS 8.6

1 Continue the example until $x = 1.0$, and compare the answer with the true solution.
2 Integrate

$$y' = y^2 + x^2 \qquad 0 \leq x \leq 1.0$$

using $h = 0.1$ and $y(0) = 0$.

8.7 Adams-Moulton method †

Although the modified Euler method is very effective, it is often preferable to use a higher-accuracy formula. A pair of such formulas is given by:

Predictor

$$y_{n+1} = y_n + \frac{h}{24} \left(55y'_n - 59y'_{n-1} + 37y'_{n-2} - 9y'_{n-3} \right) + \frac{251}{720} h^5 y^5 (\theta)$$

Corrector

$$y_{n+1} = y_n + \frac{h}{24} \left(9y'_{n+1} + 19y'_n - 5y'_{n-1} + y'_{n-2} \right) - \frac{19}{720} h^5 y^5 (\theta)$$

Thus, the **"predictor minus corrector"** is

$$\frac{270}{720} h^5 y^{(5)}$$

and if we want to **modify**, we subtract

$$\frac{251}{270} (p_n - c_n)$$

---

†See almost any standard text, for example, R. W. Hamming, "Numerical Methods for Scientists and Engineers," chap. 15, McGraw-Hill Book Company, New York, 1962, or A. Ralston, "A First Course in Numerical Analysis," chap. 5, McGraw-Hill Book Company, New York, 1965.

whereas to **final adjust**, we add

$$\frac{19}{270}\,(p_{n+1} - c_{n+1})$$

Another widely used method (which we will not derive because the derivation is not like anything else in numerical methods and therefore does not illustrate an important idea) is the Runge-Kutta method. Here we use the formulas (see the same references)

$$k_1 = f(x_n, y_n)$$

$$k_2 = f\left(x_n + \frac{h}{2},\ y_n + k_1\frac{h}{2}\right)$$

$$k_3 = f\left(x_n + \frac{h}{2},\ y_n + k_2\frac{h}{2}\right)$$

$$k_4 = f(x_n + h,\ y_n + k_3 h)$$

$$x_{n+1} = x_n + h$$

$$y_{n+1} = y_n + \frac{h}{6}\,(k_1 + 2k_2 + 2k_3 + k_4)$$

---

1 *Take the first slope k_1, and go halfway across the interval.*
2 *Use this slope k_2, and again start and go halfway.*
3 *Use this slope k_3, and go all the way.*
4 *Average k_1, k_2, k_3, and k_4, and use this as the final slope.*

This makes an excellent starting method for the Adams-Moulton method and is occasionally used to integrate an entire solution. The latter is **not** recommended because of the lack of error control and the excessive computational labor of the four function evaluations per step, but it does save programming time and effort if no library routine is available.

8.8 Step size

The starting step size to use can be found by experiment or experience or from physical intuition.

Later, the $p_{n+1} - c_{n+1}$ gives an indication of when to halve or double. Suitable halving formulas for Adams-Moulton are

$$y_{n-1/2} = \frac{1}{128} \ [45y_n + 72y_{n-1} + 11y_{n-2} + h(-9y_n' + 36_{n-1}' + 3y_{n-2}')]$$

$$y_{n-3/2} = \frac{1}{128} \ [11y_n + 72y_{n-1} + 45y_{n-2} - h(3y_n' + 36y_{n-1}' - 9y_{n-2}')]$$

For Euler's modified method we can use

$$y_{n+1/2} = \frac{y_n + y_{n+1}}{2} + \frac{h}{8} \ (y_n' - y_{n+1}')$$

When it comes to doubling, we can

1 Carry old back values.
2 Restart.
3 Use special formulas for a couple of steps.

For example, to use method 3 for Euler's predictor, we need a formula of the form

$$y_{n+2} = a_0 y_n + a_1 y_{n-1} + h(b_0 y_n' + b_1 y_{n-1}')$$

To find such a formula, we proceed pretty much as usual and make the formula exact for (because the formula is around $x = nh$):

$$y = 1, \qquad x - nh, \qquad (x - nh)^2, \qquad (x - nh)^3$$

$$
\begin{array}{lll}
y = 1 & : \ \ 1 & = a_0 + a_1 \\
y = x - nh & : \ \ 2h & = \quad - ha_1 + h(b_0 + b_1) \\
y = (x - nh)^2 : & \ \ 4h^2 = & \quad h^2 a_1 + 2h^2(-b_1) \\
y = (x - nh)^3 : & \ \ 8h^3 = & \quad - h^3 a_1 + 3h^3(b_1)
\end{array}
$$

whose solution is

$$
\begin{array}{ll}
a_0 = -27 & a_1 = 28 \\
b_0 = 18 & b_1 = 12
\end{array}
$$

8.9 Systems of equations

Although we have apparently discussed only the solution of a first-order differential equation, the methods can easily handle a system of n first-order equations

$$
\begin{aligned}
y_1' &= f_1(x, y_1, y_2, \ldots, y_n) \\
y_2' &= f_2(x, y_1, y_2, \ldots, y_n) \\
&\cdots\cdots\cdots\cdots\cdots\cdots\cdots\cdots\cdots \\
y_n' &= f_n(x, y_1, y_2, \ldots, y_n)
\end{aligned}
$$

by the simple process of doing each operation in parallel on each equation.

Higher-order equations such as

$$ y'' + y = 0 $$

are easily reduced to a system of first-order equations by the notational trick of writing

$$ y' = z $$

then

$$ y'' = z' $$

and the equation $y'' + y = 0$ is equivalent to the two equations

$$
\begin{aligned}
z' + y &= 0 \\
y' &= z
\end{aligned}
$$

Thus, our discussion really covered systems of first- and higher-order equations.

PROBLEM 8.9

1 Reduce the system

$$
\begin{aligned}
y'' &= f(x, y, y', z, z') \\
z'' &= g(x, y, y', z, z')
\end{aligned}
$$

to a system of four first-order equations.

8.10   Linear equations with constant coefficients

As a completely different approach which is based on the polynomial approximation of a coefficient in the equation and **not** the polynomial approximation of the solution, consider the equation

$$y'' + ay' + by = f(x) \qquad \begin{aligned} y(a) &= y_a \\ y'(a) &= y'_a \end{aligned}$$

Let us approximate $f(x)$ by a sequence of straight lines.

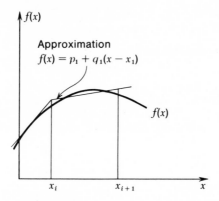

Thus, in the interval $x_i \le x \le x_{i+1}$, we assume

$$f(x) = p_i + q_i(x - x_i)$$

Given $y(x)$ and $y'(x)$ at the left of an interval, we can proceed to get the analytical solution of the equation in the interval as follows: We know the solution of the **homogeneous** equation is

$$c_1 e^{m_1(x - x_i)} + c_2 e^{m_2(x - x_i)}$$

where m_1 and m_2 are the roots of the characteristic equation

$$m^2 + am + b = 0$$

and we assume that $m_1 \ne m_2$. A particular solution of the **complete** equation can be found by trying a solution of the form

$$y = A_i + B_i(x - x_i)$$

Thus, we have

$$y' = B_i$$

$$y'' = 0$$

$$aB_i + bA_i + bB_i(x - x_i) = p_i + q_i(x - x_i)$$

Equating like terms, we have

$$bB_i = q_i$$

$$aB_i + bA_i = p_i$$

Thus,

$$B_i = \frac{q_i}{b}$$

$$A_i = \frac{p_i}{b} - \frac{q_i}{b^2}$$

and the **general solution** is

$$y(x) = C_1 e^{m_1(x-x_i)} + C_2 e^{m_2(x-x_i)} + \left(\frac{p_i}{b} - \frac{q_i}{b^2}\right) + \frac{q_i}{b}(x - x_i)$$

We can now fit the boundary conditions

$$y(x_i) = y_i = C_1 + C_2 + \frac{p_i}{b} - \frac{q_i}{b^2}$$

$$y'(x_i) = y_i' = C_1 m_1 + C_2 m_2 + \frac{q_i}{b}$$

The determinant of the unknowns C_1 and C_2 is $m_2 - m_1$ and is not zero unless the two characteristic roots are equal (we assumed they were not equal). Thus we know C_1 and C_2, and hence the required solution of the differential equation in the interval

$$x_i \leq x \leq x_{i+1}$$

We next evaluate these expressions at x_{i+1} to get the boundary conditions for the start of the next interval. With this we can repeat the cycle again and again.

Notice that we are approximating the "forcing function" $f(x)$ by a straight line in each (not necessarily equal-sized) interval.

There is no reason why we had to use straight lines; we could with slightly more trouble use quadratics or cubics if we wanted to.

Evidently, all this tedious arithmetic can be programmed for a computer.

The reason for discussing this method is to show that there are special methods which approximate the equation, or part of it, by a polynomial rather than approximating the solution as a polynomial as do most of the classical methods.

PROBLEMS   8.10

1 Discuss the case of equal characteristic roots.
2 Solve in more detail

$$y'' + y = f(x), \qquad y(0) = 1, \qquad y'(0) = 0$$

where

$$f(x) = \begin{cases} 0 & 0 \le x \le \dfrac{\pi}{4} \\[2mm] \left(x - \dfrac{\pi}{4}\right)\dfrac{4}{\pi} & \dfrac{\pi}{4} \le x \le \dfrac{\pi}{2} \\[2mm] 1 & \dfrac{\pi}{2} \le x \le \pi \end{cases}$$

8.11   Two-point problems

We have so far considered the **initial-value problem** for which all the boundary conditions are at a single (initial) point. In many applications we have boundary conditions at two (or more) points.

One widely used method of solving a **two-point problem** is that classically used in the practice of gunnery where the conditions on the trajectory of the shell are at two points, at the firing gun and at the target.

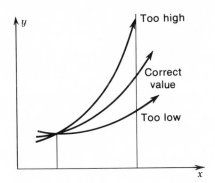

One simply guesses at an elevation for the gun, fires, notes the miss distance, adjusts the elevation, fires, notes the miss distance, and so forth.

A second method for solving a two-point problem is to divide the total interval into n subintervals, write some difference equations for each internal interval, and solve the resulting system of algebraic equations. This method is highly developed, and we shall give only one example.

Suppose we face the problem

$$y'' = f(x,y) \qquad \begin{array}{l} y(0) = 0 \\ y(1) = 0 \end{array}$$

We arbitrarily take a spacing of $h = 0.2$ and have

$$y_{k+1} - 2y_k + y_{k-1} = h^2 f_k = h^2 f(x_k, y_k)$$

$$\text{for } k = 1, 2, 3, 4$$
$$x_k = 0 + 0.2k$$

These equations in more detail are

$$\begin{aligned} y_2 - 2y_1 &= h^2 f_1 \\ y_3 - 2y_2 + y_1 &= h^2 f_2 \\ y_4 - 2y_3 + y_2 &= h^2 f_3 \\ -2y_4 + y_3 &= h^2 f_4 \end{aligned}$$

where $f_i = f(x_i, y_i)$.

They form a set of four equations in the four unknowns y_1, y_2, y_3, and y_4.

If the term

$$\left| h^2 \frac{\partial f}{\partial y} \right| < 1$$

then we can hope that an iterative solution will work.

As a special case, consider

$$f(x,y) = xy^2 + 1$$

We need to guess at a solution for a first trial. Let us simply take $y = 0$. Then,

$$\begin{aligned} y_2 - 2y_1 &= h^2 \\ y_3 - 2y_2 + y_1 &= h^2 \\ y_4 - 2y_3 + y_2 &= h^2 \\ -2y_4 + y_3 &= h^2 \end{aligned}$$

Adding gives $-(y_4 + y_1) = 4h^2$, and the initial symmetry shows

$$y_4 = y_1 = -2h^2 = -0.08$$
$$y_2 = h^2 + 2y_1 = -3h^2 = y_3 = -0.12$$

We guessed $y \equiv 0$; now we can use these calculated values to compute some new f_i

$$f_1 = 1.001$$
$$f_2 = 1.058$$
$$f_3 = 1.086$$
$$f_4 = 1.005$$

It should be clear that we can use these values as our new starting guesses and that a repetition of the cycle will converge rapidly to the solution.

If the original differential equation had been linear, then the equations to be solved would be linear and an iterative approach would not be necessary.

OPTIMIZATION 9

9.1 Introduction

Much of the complex art of engineering is the art of optimization. Therefore, it should be no surprise to the student to be told that in practice optimization can be very difficult, and only special cases can be handled easily and reliably.

Optimization is the third stage in the process of designing a system. The first stage is modeling, or simulating, the system. This stage includes the art and special knowledge of the field of application. Upon the quality of the modeling depends the value of the answer. All too often a poor model is accurately optimized; usually it is wiser to optimize a better model partially.

System Design

Step 1   *Construct model.*

Step 2   *Construct objective function.*

Step 3   *Optimize.*

The second step is to decide what is to be optimized, that is, to construct the so-called objective function that describes what is to be optimized (often the cost is to be minimized). This construction again lies in the field of application and outside the domain of a book on numerical methods. The construction of the objective function is often so difficult that in the past it has been left in a vague state of a general understanding of what is to be done, and this has at times led to poor workmanship because the designers did not clearly recognize what they were trying to do. However, let us be realistic and admit that in many applications it is not possible to say at the beginning just exactly what is to be optimized; often it is only after the matter has been examined from many sides that the true nature of the problem is understood to any extent. When we suppose that we are given an objective function to optimize, we are in fact supposing that the problem is well understood and that the function we are given does indeed describe the optimization problem accurately.

Furthermore, in many cases of engineering design and evaluation, there are many goals to be considered. In these cases the optimizations for each of the goals must be combined at some time, either analytically through some new objective function that combines the various goals, or by some intuitive judgment. It is surprising how after long and careful analytical studies the final judgment is frankly a "hunch."

Optimization implies either maximizing or minimizing. But since the maximum of a function $f(x)$ occurs at the same place as does the minimum of $-f(x)$, and the extreme values are simply related, it is convenient to discuss only the minimization problem; this we will do.

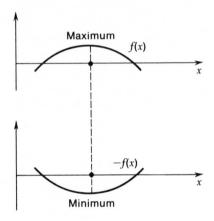

The classical design methods often used the simple process of setting up the model plus the objective function (if only vaguely in the mind) and then adjusting one variable at a time in trying to minimize the numerical value of the objective function. In simple problems this may be satisfactory, but in more complex problems this method of "one variable at a time" is too slow and costly for practical use, and we now have methods which change many or all of the variables at the same time.

Let us emphasize again, optimization is implied in almost every simulation or modeling, but the success of the optimization depends critically on the quality of the modeling and on the proper choice of the objective function. If either is not accurate or reasonably close to accurate, then it is highly unlikely that the result of a careful optimization will be of much value. Thus, before starting any optimization a careful examination should be made of the previous two steps.

9.2   Results from the calculus

We begin with a brief review of optimization as it is usually taught in the calculus course. In a calculus course the function characteristically has a horizontal tangent at the point where a maximum or minimum occurs. For a function of one variable this means that the derivative is zero at the point, whereas for a function of many variables it means that all the partial derivatives at the point are zero. In both cases this is not sufficient; it is only necessary.

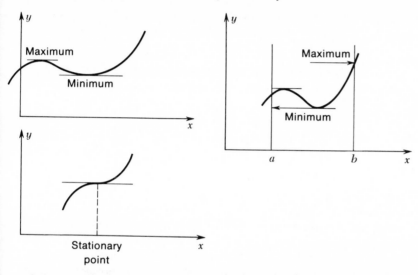

It is usually glossed over that the interval to be searched for extremes may be limited and that a special check needs to be made along the boundary. This remark will serve to recall that in fact what the method of "equating the derivative to zero" really found was **relative** maxima or minima, and that sometimes points which were found were neither; these are the so-called **stationary points** where the function is locally horizontal but where there are both larger and smaller values in the neighborhood. Furthermore, singularities, such as cusps in the function, were tacitly excluded.

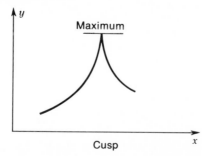

For functions of a single variable, the additional test with the sign of the second derivative (supposing that the value of the second derivative is not zero) will separate the maxima and minima from each other as well as exclude the stationary points.

Second derivative test

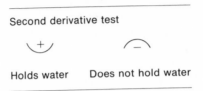

Holds water   Does not hold water

For functions of more than one variable, the textbooks rarely go beyond two variables and only observe that

$$f_{xx}f_{yy} - f^2_{xy} \begin{cases} > 0 & \begin{cases} f_{xx} < 0 & \text{max} \\ f_{xx} > 0 & \text{min} \end{cases} \\ = 0 & \text{undecided} \\ < 0 & \text{neither max nor min} \end{cases}$$

In all this it is tacitly assumed that the simultaneous equations that occur can be solved. For a single variable the methods in Chap. 2 can be used, but for several variables there are no really good methods for solving the equations that give the locations of the possible extreme values, although some methods use what amounts to Newton's method applied to many variables. Since the extremes are among the zeros of the derivative, all the derivatives used in an optimization are one higher than the corresponding derivatives in a zero-finding process.

An example

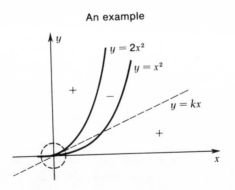

Consider

$$z = (y - x^2)(y - 2x^2)$$
$$= y^2 - 3x^2 y + 2x^4$$

At origin,

$$\frac{\partial x}{\partial x} = -6xy + 8x^3 = 0$$

$$\frac{\partial z}{\partial y} = 2y - 3x^2 = 0$$

$$f_{xx} f_{yy} - f_{xy}{}^2 = 0$$

But note that any plane

$$y = kx$$

through the origin cuts the surface in

$$z = k^2 x^2 - 3kx^3 + 2x^4$$

$$\frac{\partial z}{\partial x} = 2k^2 x - 9kx^2 + 8x^3 = 0$$

$$\frac{\partial^2 z}{\partial x^2} = 2k^2 > 0$$

True relative minima at origin, **but** *every neighborhood of origin has negative values!*

Another warning needs to be given, this one illustrating that the extreme value may not exist. Suppose you are in town A and wish to go to town B which is due east, but you are required to depart from A in the direction of north. What is the shortest path? Evidently there is none in a strict mathematical sense.

A B

PROBLEMS 9.2

1 Find the maxima and minima of

$$y = 1 - 6x^2 + 4x^3 \qquad (-2 \leq x \leq 2)$$

2 Find the extremes of

$$y = x^3 \qquad (-1 \leq x \leq 1)$$

3 Find the maxima ($0 \le x \le \infty$) of

$$y = x^k e^{-x}$$

4 If sufficient derivatives are supposed to be available, characterize maxima, minima, and stationary points in one independent variable.

5 Find the maxima of

$$z = -1 + 2x - 2y - x^2 - y^2$$

9.3 The gradient

The methods of the calculus try to go directly from the function to the desired value, or values. The iterative methods we shall now examine are based on the simple principle of going downhill to find the lowest point. These methods take their inspiration from the topographic map, which shows curves of equal altitude.

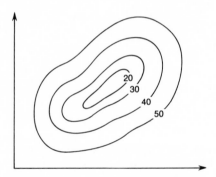

In a similar way we consider the **level curves** of constant objective-function value. Such curves are defined by

$$f(x_1, x_2, \ldots, x_n) = \text{constant}$$

As a general observation on the weakness of going downhill to find the lowest point, consider the situation in which there is a landscape of rolling hills and also a deep mine-shaft opening on the side of a hill.

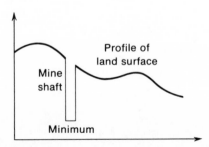

Unless by chance one found the mine-shaft opening, it is unlikely that he would find the lowest place. About all that can be done in the general case is to start at a number of well-chosen places, then go downhill from them until there is no lower place to go, and each time find a local minimum. We then select among the local minima the one that is lowest and take that as the minimum, realizing full well that we may have completely missed better places.

In more detail, let the objective function be

$$y = f(x_1, x_2, \ldots, x_n) = \text{constant}$$

For y equal to a constant, we have level curves (surfaces and hypersurfaces for $n > 2$) in the x_i variables. These level curves will typically enclose a local minimum. If we imagine differentiating the objective function while holding the value of y constant for a level curve, we get

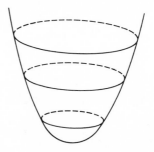

$$\frac{dy}{ds} = 0 = \frac{\partial f}{\partial x_1} \frac{\partial x_1}{\partial s} + \frac{\partial f}{\partial x_2} \frac{\partial x_2}{\partial s} + \cdots + \frac{\partial f}{\partial x_n} \frac{\partial x_n}{\partial s}$$

where s is the arc length along the level curve. The n functions

$$\frac{\partial x_j}{\partial s} \qquad (j = 1, 2, \ldots, n)$$

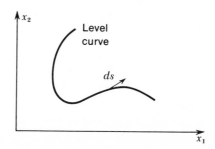

are the direction cosines of the level curve; hence we know that the

$$\frac{\partial f}{\partial x_j} \qquad (j = 1, 2, \ldots, n)$$

are direction numbers (proportional to the direction cosines) that point in the direction perpendicular to the level curve. They can be regarded as components of a column vector which we shall indicate by boldface type, $\partial \mathbf{f}/\partial \mathbf{x}_j$.

The gradient is the column vector

$$\begin{pmatrix} \dfrac{\partial f}{\partial x_i} \\[2mm] \dfrac{\partial f}{\partial x_i} \\[1mm] \vdots \\[1mm] \dfrac{\partial f}{\partial x_i} \end{pmatrix} \equiv \frac{\partial \mathbf{f}}{\partial \mathbf{x}_i}$$

These numbers serve to define the local direction of the **gradient** which can be shown to be the direction of steepest ascent (locally), the negative gradient being therefore the direction of steepest descent.

The gradient will, of course, change its direction from point to point on the objective-function surface, but if we could follow the direction of the local negative gradient we should be led to the local minimum.

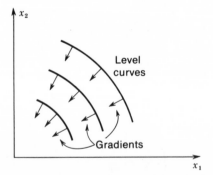

Since all the nearby negative gradient paths lead to the same local minimum, it is not necessary to follow the negative gradient exactly; we can be rather careless and still expect to get to the local minimum.

The simple gradient method has the weakness that near the minimum the negative gradient points only very weakly (because of round-off, if nothing else) to the minimum.

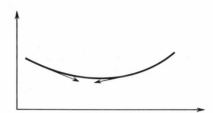

However, this weakness should not be exagerated; if all the slopes are very small, then there is probably very little to be gained by further refinement of the estimate of the position of the minimum. Very likely the model of the situation is not sufficiently accurate to justify great precision in this matter.

Evidently, the gradient method resembles the direction-field approach to the solution of differential equations. In principle we need to follow the trajectory defined by the local negative gradient.

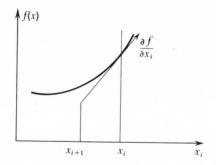

The crudest method is to take steps [we use the superscript (i) in parentheses to indicate the iteration step number]

$$x_j^{(i+1)} = x_j^{(i)} - h \frac{\partial f^{(i)}}{\partial x_j} \qquad (j = 1, \ldots, n)$$

where h is the step size. The first problem is to guess at a reasonable step size h, with the realization that **we do not need to be accurate** since all nearby paths lead to the same local minimum. One way of monitoring the step size is to compute at each step

$$\sum_{j=1}^{n} (x_j^{(i+1)} - x_j^{(i)})^2$$

and if this is too small, then increase the step size. A more sensible method would be to use

$$\left[\sum_{j=1}^{n} \left(\frac{\partial f^{(i)}}{\partial x_j} \right)^2 \right]^{1/2}$$

to convert the direction numbers into direction cosines. In this way we keep about the same amount of progress at each step.

The fault of this method is simply that for a reasonably high dimensional space the gradient

$$\frac{\partial \mathbf{f}^{(i)}}{\partial \mathbf{x}_j} \qquad (j = 1, 2, \ldots, n)$$

takes a great deal of computing time, and it is probably not worth the effort to follow the local gradient even approximately. However, if you decide to follow the negative gradient down to the local minimum by integrating the differential equations (where s is the arc length in the $\mathbf{x}_j$ variable space)

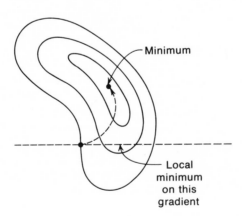

and then perhaps the modified Euler method would be better than the crude method. Limited experience seems to suggest that the whole approach is not efficient in machine time used.

$$\frac{\partial \mathbf{x}_j^{(i)}}{\partial \mathbf{s}} = - \frac{\partial \mathbf{f}}{\partial \mathbf{x}_j}$$

then perhaps the modified Euler method would be better than the crude method. Limited experience seems to suggest that the whole approach is not efficient in machine time used.

9.4   Some practical remarks

The valleys that seem to be the cause of most of the trouble are long, thin, curving ones and are of more frequent occurrence than might at first be thought.

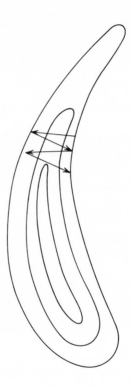

For such valleys the successive steps may well oscillate back and forth across the valley. The same idea of not taking a step that increases the minimum that was applied in Bairstow's method (Chap. 4) seems relevant here. We shall examine this under the name, "the method of steepest descent" (Sec. 9.6).

We have in the past pointed out the importance of preliminary scaling of the problem, and this idea applies here also; we should in some sense use comparable units in the various x_i, but again, as in the past, this is easier said than done. But evidently, if we had a very flattened elliptically shaped valley, we should expect more trouble with our method than if we made the appropriate change of scale and transformed it into a circular valley.

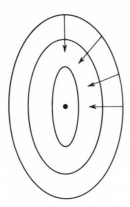

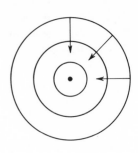

In a circular valley the negative gradient always points toward the minimum.

The last remark points out a weakness of our previous methods; it would seem that a really good method would be independent of scaling. If we ask for a method that will go in the right direction for elliptical valleys, we can indeed find one, such as the variable metric method to be described later (Sec. 9.7).

Some of the more highly favored methods in the literature will switch to a variable metric method when they are close to the minimum in the belief that in the immediate neighborhood of a minimum the behavior of the surface is very close to a conic (in n variables, of course).

9.5 Estimating the gradient

In many problems it is either difficult or impossible to get an expression for the components of the gradient in a closed form. For example, the function values may be the results of experiments. In such cases the simple expedient of changing one coordinate a small amount and observing the corresponding change in the function will give an estimate of the corresponding partial derivative.

$$\frac{f(x + \Delta x) - f(x)}{\Delta x} \simeq \frac{\partial f}{\partial x}$$

The weakness of this process is that if the change in the coordinate is small, then the corresponding difference in the function is also small and the estimate of the derivative is very subject to roundoff noise (a common error); whereas if the change is large, then it no longer measures the **local** gradient component. Still, the method is sometimes used and is often moderately effective. When this simple modification is used, a change in each coordinate to get each of the local derivative estimates is required to get the gradient, and this may be an exorbitant price to pay.

9.6 Steepest descent

The negative gradient

$$-\frac{\partial \mathbf{f}^{(i)}}{\partial \mathbf{x}_j}$$

at the ith step points in the direction of steepest descent because it is perpendicular to the local level curve. We propose to search along this direction for a local minimum. To do this, we step forward, using equal-sized steps, until we find three equally spaced points for which the objective function $f(x_1, \ldots, x_n)$ has a lower value at the middle point than it has at the other two points. If on the first step the value of the objective function rises, then we simply bisect and examine it again. Repeated bisections, if necessary, will ultimately yield a triple of equally spaced points (until roundoff interferes), because we know we are starting in a downward direction.

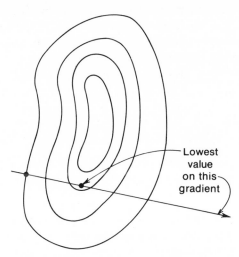

Lowest value on this gradient

Having found the three points for which the middle has the lowest value of the objective function, we use quadratic interpolation to find an estimate of the minimum. There is little point in being very accurate in the early stages of the process.

At this minimum point on the line, we should be (if we were exactly correct, which we are not) at the point where the line along which we have been searching is tangent to the local level curve. Thus, at this point the new gradient will be perpendicular to the old gradient we were using; our next search direction will be orthogonal to the present one.

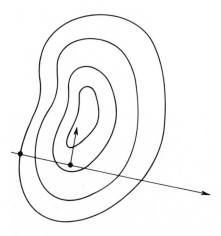

By repeating this search pattern each time down the negative gradient, until we come to a minimum on the line pointing in that direction and then searching in the new direction of steepest descent, we shall approach a local minimum of the surface of the objective function. In this process we evaluate the function frequently, but we avoid the frequent evaluation of the gradient.

When shall we stop the process? There is no completely satisfactory rule for stopping, but the following two are quite popular:

$$\sum_{j=1}^{n} \left(\frac{\partial f^{(i)}}{\partial x_j} \right)^2 < \varepsilon$$

$$\sum_{j=1}^{n} \left| f\left(x_j{}^{i+1}\right) - f\left(x_j{}^{i}\right) \right| < \varepsilon$$

where $\mathbf{x}^{(i)}$ is the vector of the ith iteration. Both have their faults.

In more detail, the method of steepest descent consists of the following steps:

1   Start at some suitable initial point $\mathbf{x}^{(0)}$ (a point having n coordinates). The ith iteration proceeds as follows.

2 Compute the negative gradient.

$$-\frac{\partial \mathbf{f}^{(i)}}{\partial \mathbf{x}_j}\bigg|_{\text{at } \mathbf{x}=\mathbf{x}^{(i)}} \quad \text{direction}$$

The steps to be taken lie in this direction, that is,

$$\mathbf{x}_j^{(i-1)} = \mathbf{x}_j^{(i)} - h\,\frac{\partial \mathbf{f}^{(i)}}{\partial \mathbf{x}_j}$$

where h is a parameter.

3 Search in the negative gradient direction until three points are found for which the middle one has the least of the three values of the objective function. This can be done by stepping forward in equal-sized steps until a larger value than the immediately preceding one is reached. If the first step has this property, we bisect and try again, repeating the bisection until we find three equally spaced values with the desired property.

4 Use quadratic interpolation about the middle point

$$t = \frac{-h}{2}\left(\frac{\mathbf{f}_h^{(i)} - \mathbf{f}_{-h}^{(i)}}{\mathbf{f}_h^{(i)} - 2\mathbf{f}_0^{(i)} + \mathbf{f}_{-h}^{(i)}}\right)$$

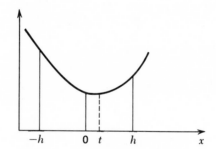

to find the minimum along the line that was being searched, where

$$\mathbf{f}_h^{(i)} = f\left(\mathbf{x}_j^{(i)} - h\,\frac{\partial \mathbf{f}^{(i)}}{\partial \mathbf{x}_j}\right)$$

Thus, the new point is

$$\mathbf{x}_j^{(i+1)} = \mathbf{x}_j^{(i)} - t\,\frac{\partial \mathbf{f}^{(i)}}{\partial \mathbf{x}_j}$$

for each j.

5 Check to see if the stop criterion is met, and if not, return to step 2. Notice that as the iterations progress, it will usually be necessary to decrease the search-step size h.

9.7 A variable metric method

Most variable metric methods can be considered quasi-Newton methods. In Newton's method for finding a zero of a function of a single variable, we use both the function and the first derivative values. In searching for a minimum (a zero of the derivative), we should use the derivative and the second derivative.

$$-\frac{f'(a)}{f''(a)}$$

For many variables we should need the inverse of the matrix of second partial derivatives (the Hessian matrix).

Example

Let $f(x)$ be a quadratic

$$f(x) = a + b^T x + x^T Q x$$

Then the gradient is

$$g(x) = b + Qx$$

Let $x°$ be the minimum and x^0 any initial point. Then,

$$g(x°) = g(x^0) + Q(x° - x^0)$$

But $g(x°) = 0$; hence

$$x° = x^0 - Q^{-1} g(x^0)$$

and if Q^{-1} is known, then the minimum is found in one step.

If we had this inverse, and if the surface were exactly a quadratic, then we should expect to find the minimum in one step. In practice we neither evaluate the Hessian (since it is usually very expensive to do so), nor find the inverse; furthermore, the surface that we are searching for a minimum is not apt to resemble a quadratic until we are quite near the minimum.

The variable metric, or quasi-Newton, methods all have the underlying idea of generating the direction of the next step of the search by multiplying the negative gradient by a matrix that in some sense is related to the inverse Hessian. Corrections in this matrix are later made so that it approaches the inverse Hessian. Perhaps the best known of these methods is the one made popular by Fletcher and Powell.[†]

[†] R. Fletcher and M. J. D. Powell, "A Rapidly Converging Descent Method for Minimization, *Brit Comp. J.*, vol 6, pp. 163–168, 1963.

The steps are as follows:

1 Start with the matrix I as the initial guess $H^{(0)}$ for the matrix of quadratic terms. For the ith steps we proceed as follows:

2 Compute the negative gradient just as before,

$$-\frac{\partial f^{(i)}}{\partial x_j}$$

3 Compute the new direction

$$\mathbf{s}_j^{(i)} = -H_i^{(j)}\left(-\frac{\partial f}{\partial x_j}\right)$$

4 Find the step size t as before by quadratic interpolation.

5 Compute the vector

$$\boldsymbol{\sigma}_j^{(i)} = t\mathbf{s}_j^{(i)}$$

6 Compute the new value as before,

$$x_j^{(i+1)} = x_j^{(i)} + ts_j^{(i)}$$

7 Compute the change in the negative gradient,

$$\mathbf{y}_j^{(i)} = \left(-\frac{\partial f^{(i+1)}}{\partial x_j}\right) - \left(-\frac{\partial f^{(i)}}{\partial x_j}\right)$$

8 Compute the matrix

$$A^{(i)} = \frac{\boldsymbol{\sigma}_j^{(i)}\,\boldsymbol{\sigma}_j^{(i)T}}{\boldsymbol{\sigma}_j^{(i)T}\,\mathbf{y}_j^{(i)}}$$

where the superscript T means the transpose.

9 Compute the matrix

$$B^{(i)} = -\frac{h^{(i)}\mathbf{y}^{(i)}\mathbf{y}^{(i)T}H^{(i)}}{\mathbf{y}^{(i)T}H^{(i)}\mathbf{y}^{(i)}}$$

10 Compute the matrix for the next iteration

$$H^{(i+1)} = H^{(i)} + A^{(i)} + B^{(i)}$$

11 Check the stop criterion, and if it is not satisfied, then return to step 2 for the next iteration.

9.8 Optimization subject to linear constraints

Up to now the problems in this chapter have been problems of unconstrained optimization. This has meant that in our search for a minimum of $f(x_1, \ldots, x_n)$, any value of the variables $x_1, \ldots, x_n$ was a permissible one. There are many practical cases, however, in which physical or mathematical reasoning forces us to restrict our search to values of the variables which satisfy certain conditions.

Minimize

$$f(x) = (x_1 - 4)^2 + (x_2 - 4)^2$$

subject to

$$x_2 - x_1 \geq 0$$
$$4x_1 + x_2 + 12 \geq 0$$
$$-x_1 - x_2 + 4 \geq 0$$
$$x_1 - x_2 + 5 \geq 0$$

These conditions, or constraints as we will call them from now on, can take on many forms. In this section and the next we shall examine some of the forms in which constraints appear and how they effect the solution and we shall suggest methods for handling them.

Some constraints are easier to handle than others. Consider, for example, nonnegative constraints, that is, each of the variables $x_1, \ldots, x_n$ is restricted to nonnegative values. Under these conditions, the statement of the optimization problem would read: Minimize the function

$$f(x_1, \ldots, x_n)$$

subject to the constraints

$$x_i \geq 0 \qquad (i = 1, \ldots, n)$$

The easiest way to solve this problem is through a transformation of variables. If we let

$$x_i = y_i^2 \qquad (i = 1, \ldots, n)$$

then we may minimize f as an unconstrained function of the $y_1, \ldots, y_n$ and be certain that the nonnegative constraints will be satisfied at the optimum. An alternative way is to apply the gradient or the quasi-Newton technique of Sec. 9.7 with the following modification: Whenever a variable, say x_k, becomes zero and at the same time the

xth component of the gradient is positive, then stop updating x_k until $\partial f/\partial x_k$ changes sign. This is another way in which we can be sure that the search will be restricted to nonnegative values.

The above tricks apply equally well when the variables are restricted to lying between lower and upper bounds, that is, when the constraints are of the form

$$x_i - l_i \geq 0$$
$$u_i - x_i \geq 0 \qquad (i = 1, \ldots, n)$$

where the l_i and u_i are given and, of course, $u_i > l_i$. Here, we may again use a transformation on the variables. We may, for example, let

$$x_i = (u_i - l_i) \sin^2 y_i + l_i$$

and solve the problem as unconstrained in terms of the y's. Other transformations are possible; however, one has to be careful about introducing new local minima. This is because the minimization of $f(y_1, \ldots, y_n)$ will stop whenever $\partial f/\partial y_i = 0$; $i = 1, \ldots, n$. But by the chain rule,

$$\frac{\partial f}{\partial y_i} = \frac{\partial f}{\partial x_i} \frac{\partial x_i}{\partial y_i}$$

Therefore, any transformation on the x's which causes $\partial x_i/\partial y_i$ to be zero at a point which meets this constraint and is not on the boundary will introduce a new minima.

The constraint types discussed so far are linear and are special cases of the general linear constraints of the form

$$\sum_{i=1}^{n} a_{ij} x_i - b_j \geq 0 \qquad (j = 1, \ldots, m)$$

Linear constraints are much easier to handle than the nonlinear ones, which we discuss in the next section. The main reason for this is that the boundary of the feasible region is composed of straight lines in two dimensions, planes in three dimensions, and hyperplanes in higher dimensions. It is, consequently, easy to travel along a boundary during the search for the optimum. Several methods, developed especially for linearly constrained problems, take full advantage of this fact. Best known among these is the so-called gradient-projection method. This method has been designed primarily for linearly constrained problems with a nonlinear objective function; it is essentially an extension of the steepest-descent method to linear constraints. This presentation is too lengthy to be included in this book and the reader is referred to the literature.†

A very important class of optimization problems is the one in which both the objective function and all the constraints are linear. Problems of this type, that is, minimize

$$f(x) = \sum_{i=1}^{n} c_i x_i$$

subject to

$$\sum_{i=1}^{n} a_{ij} x_i - b_j \geq 0 \qquad (j = 1, \ldots, m)$$

are called **linear programming** problems, and they are important because of the enormous variety of applications which they have found in practical situations.‡ A little thought will make it intuitively clear that the solution of a linear program lies on the boundary of the feasible region, usually on a corner, but sometimes along one side of a constraint. This observation provides an efficient technique for solving linear programs. More details can, again, be found in the literature.

9.9 Nonlinear constraints—Lagrange multipliers

Up to now the constraints we have used have been linear inequalities. Had they been equalities, it would have been easy (in principle) to eliminate some of the variables and thus reduce the problem to one with no linear equality constraints.

A constraint that is a nonlinear function of the variables makes the direct elimination appear to be difficult. Fortunately, the method of Lagrange multipliers makes it fairly easy in theory to handle the problem.

Suppose we wish to find the stationary values (the extreme values plus possibly a few others) of the function

$$z = f(x,y)$$

---

†Read, for example, the section beginning on p. 133 of "Nonlinear Mathematics" by T. L. Saaty and J. Bram, McGraw Hill Book Company, New York, 1964.

---

‡See, for example, S. I. Gass, "Linear Programming," 3d ed., McGraw-Hill Book Company, 1969.

subject to the constraint

$$g(x,y) = 0$$

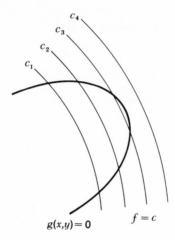

$$g(x,y) = 0 \qquad f = c$$

The proof is straightforward. Suppose for the moment that from $g(x,y) = 0$ we find

$$y = \phi(x)$$

Then we want to find the stationary values of

$$f(x, \phi(x))$$

Thus, we differentiate f and set it equal to zero

$$\frac{\partial f}{\partial x} = \frac{\partial f}{\partial x} + \frac{\partial f}{\partial y}\frac{d\phi}{dx} = 0 \qquad\qquad (9.1)$$

But $g(x,y) = 0$ leads to

$$\frac{\partial g}{\partial x} + \frac{\partial g}{\partial y}\frac{dy}{dx} = 0$$

or

$$\frac{\partial g}{\partial x} + \frac{\partial g}{\partial y}\frac{d\phi}{dx} = 0 \qquad\qquad (9.2)$$

Eliminating $d\phi/dx$ from Eqs. (9.1) and (9.2), we get

$$\frac{\partial f}{\partial x}\frac{\partial g}{\partial y} = \frac{\partial f}{\partial y}\frac{\partial g}{\partial x} \qquad (9.3)$$

We now define λ by

$$\frac{\partial f}{\partial y} + \lambda \frac{\partial g}{\partial y} = 0 \qquad (9.4)$$

and use this in Eq. (9.3) to eliminate $\partial g/\partial y$ so that we get

$$\frac{\partial f}{\partial x}\frac{\partial f}{\partial y} + \lambda \frac{\partial f}{\partial y}\frac{\partial g}{\partial x} = 0$$

or

$$\frac{\partial f}{\partial x} + \lambda \frac{\partial g}{\partial x} = 0 \qquad \left(\frac{\partial f}{\partial y} \neq 0\right) \qquad (9.5)$$

These two equations (9.4) and (9.5) together with $g(x,y) = 0$ determine the quantities λ, x, and y at the stationary point.

These equations can be obtained **formally** by the following method: In place of $f(x,y)$, we use

$$F(x,y) = f(x,y) + \lambda g(x,y)$$

The derivatives

$$\frac{\partial F}{\partial x} = \frac{\partial f}{\partial x} + \lambda \frac{\partial g}{\partial x} = 0$$

$$\frac{\partial F}{\partial y} = \frac{\partial f}{\partial y} + \lambda \frac{\partial g}{\partial y} = 0$$

lead to the same equations as before and together with the constraint $g(x,y) = 0$ determine the solution.

For n variables $x_1, x_2, \ldots, x_n$ and the function

$$f = f(x_1, x_2, \ldots, x_n)$$

together with the m constraints

$$g_1(x_1, x_2, \ldots, x_n) = 0$$
$$\cdots\cdots\cdots\cdots\cdots\cdots\cdots\cdots\cdots$$
$$g_m(x_1, x_2, \ldots, x_n) = 0$$

leads to the function

$$F = f + \lambda g_1 + \lambda_2 g_2 + \cdots + \lambda_m g_m$$

The n partial derivatives

$$\frac{\partial F}{\partial x_i} = 0 \qquad (i = 1, \ldots, n)$$

plus the m constraints

$$g_j = 0 \qquad (j = 1, \ldots, m)$$

determine the $m + n$ unknown x_i and λ_j. These equations must be satisfied at every extreme value of f unless all the Jacobians of the m functions g_j with respect to every set of m variables chosen from $x_1, x_2, \ldots, x_n$ is zero.

9.10   Other methods

The Lagrange multiplier approach is a classical one and has endured a long time because of its practical and theoretical importance. There are occasions, however, when the solution of the nonlinear equations resulting from this approach is quite difficult. In such cases, other methods are preferred in practice.

One of the best-known alternative techniques for handling equality constraints was originated by R. Courant. In this method we minimize the function

$$F(x_1, \ldots, x_n) = f(x_1, \ldots, x_n) + r \sum_{i=1}^{m} [g_i(x_1, \ldots, x_n)]^2 \qquad (r > 0)$$

as an unconstrained problem for a sequence of monotonically increasing values of r. As r becomes infinitely large, the sum of the squares of the constraints is forced to zero. In this way, the sequence of the successive minima of F converges toward the solution of the given problem.

We conclude this chapter with a few brief comments about nonlinear inequality constraints of the form

$$g_i(x_1, \ldots, x_n) \geq 0 \qquad (i \ldots 1, \ldots, m)$$

First let us examine the necessary conditions for the solution. It turns out that here, as for the equality constraints, necessary conditions can be expressed through a Lagrange function. If a point $x_1, \ldots, x_n$ is a minimum of the given objective function $f(x_1, \ldots, x_n)$ subject to the above constraints, then it must be true that the negative gradient vector of f can be expressed as a linear combination of the negative gradients of the constraints which are binding. Geometrically this means that if the minimum point is on the boundary of the feasible region, then the negative gradient of f must fall within the cone which is formed by the gradients of the constraints which are binding.

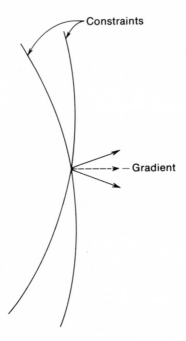

A little thought will show that if this were not true, then we should be able to move a little farther along the boundary to obtain a point at which f is smaller.

There are two common philosophies in the various solution approaches suggested for problems with nonlinear inequality constraints: "the boundary-following approaches" and the "penalty-function techniques."

As their name implies, the boundary-following approaches suggest that when a constraint is (or is about to be) violated, then follow the boundary of the feasible region defined by that constraint until a point which satisfies the necessary conditions for a minimum is found. If the boundary is highly nonlinear, then the convergence of such approaches will be slow. In such cases, we prefer the penalty-function techniques.

The idea of a penalty function was introduced earlier when we added the sum of the squares of equality constraints to f in order to "penalize" the minimization of the resulting function whenever a constraint is violated. A similar technique can be used for inequality constraints. Consider, for example, the functions

$$F(x_1, \ldots, x_n) = f(x_1, \ldots, x_n) + r \sum_{i=1}^{m} \frac{1}{g_1(x_1, \ldots, x_n)} \qquad \begin{array}{c} (r > 0) \\ r \to 0 \end{array}$$

or

$$F(x_1, \ldots, x_n) = f(\quad) - r \sum_{i=1}^{m} \ln\left[g_i(x_1, \ldots, x_n)\right] \qquad (r \to 0)$$

or

$$F(\quad) = f(\quad) + r \sum_{i=1}^{m} \left[\widetilde{g}_i(x, \ldots, x_n)\right]^2 \qquad (r \to \infty)$$

where $\widetilde{g}_i = 0$ if $g_i \geq 0$, and $\widetilde{g}_i = g_i$ if $g_i < 0$.

If we minimize these functions sequentially for a series of positive values for r (monotonically decreasing for the first two functions and increasing for the third one), then we approach a solution to the constrained problem.

10.1 The idea of least squares

Situations frequently arise in engineering and science in which there are more conditions to be satisfied than there are parameters to adjust. For example, we may be given a family of curves with, say, three parameters, and we are required to find that member of the family which best approximates a set of $M > 3$ data points. The question is, of course: Best approximates in what sense? How shall we choose the particular member of the family?

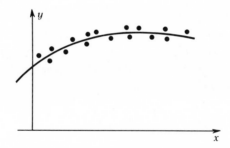

One widely used method for selecting the values of the parameters that define the particular curve of best fit is to say they are given by the choice which minimizes the sum of the squares of the differences between the observations and the chosen curve at the corresponding points.

Minimize

$$\sum_{k=1}^{M} (f_{\text{obs}} - f_{\text{approx}})^2$$

This method is called the "least-squares method of fitting the curve." Thus we have a special case of the general minimization problem which was treated in Chap. 9. The particular method of least squares has special features which deserve special treatment.

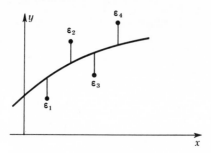

Before accepting this method as worth using, let us examine some of its features. It has been said that scientists and engineers believe that the principle of least squares is a mathematical theorem, whereas mathematicians believe it to be a physical law. Neither belief is correct; it is simply a convenient criterion for selecting a particular curve to fit some given data. The main characteristic of the method of least squares is that it puts great emphasis on large errors, and very little emphasis on small errors; the method will prefer 10 small errors of size 1 to one error of size 4.

$$4^2 > \sum_{k=1}^{10} (1)^2 = 10$$

An obvious fault of the method is that a gross blunder in the recording of the data will usually completely dominate the result.

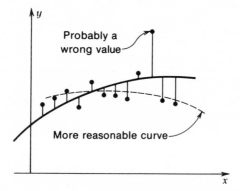

It is wise, therefore, to examine the residuals—the differences between the observed data and the computer "data" from the approximating least-squares curve—to see if they appear to be reasonable or due to a single large error (or perhaps two large errors).

Other choices of a criterion for selecting the particular member of the family of curves can be made, and of course different criteria give different answers, sometimes quite different! The least-squares choice minimizes

$$\sum_{i=1}^{M} \varepsilon_i^2$$

but we might, for example, try to minimize

$$\sum_{i=1}^{M} |\varepsilon_i|$$

where the ε_i are the residuals. This minimization of the sum of the absolute values of the residuals is rarely used, probably because it leads to difficult mathematics. Another criterion, which we shall examine more closely in Chap. 13, is

$$\min \left(\max_{i \,=\, 1, \,...,\, M} |\varepsilon_i| \right)$$

which minimizes the maximum residual error (ε_i).

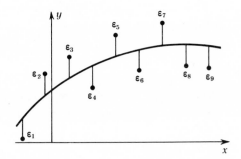

In this chapter we shall confine ourselves to the least-squares criterion.

Least-squares fitting is often regarded as a method of "smoothing the data" to eliminate unwanted "noise" or errors in the observations.

Points on the least-squares curve are often regarded as the "smoothed values" of the original data.

From the assumption that the errors satisfy certain plausible conditions, we can derive the least-squares criterion for selecting the parameters, but we shall not do so here.†

---

†R. W. Hamming, "Numerical Methods for Scientists and Engineers," chap. 17, McGraw-Hill Book Company, New York, 1962.

10.2   The special case of a straight line

In order to approach the general method, let us start with the simple and very common problem of fitting a straight line

$$y = ax + b$$

to a set of data points

$$(x_i, y_i) \qquad (i = 1, \ldots, M)$$

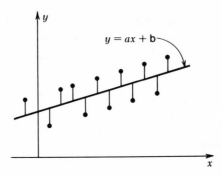

We want to pick two parameters a and b so that the calculated values

$$y(x_i) = ax_i + b$$

are near the data points y_i.

Note that y_i as used here means the data points (observations), whereas $y(x_i)$ are the calculated values

---

An example

Observations: $(0, 0)\,(1, 1)\,(2, 1)\,(3, 4)\,(4, 4)$

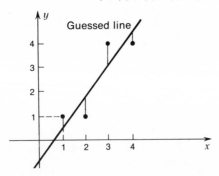

It is unreasonable to expect that the line will go exactly through each data point (x_i, y_i). Thus, we consider the **residuals**

$$\varepsilon_i = y_i - y(x_i)$$
$$= y_i - (ax_i + b)$$

as a measure of the *error*. The least-squares criterion requires minimizing

$$m(a,b) = \sum_{i=1}^{M} \varepsilon_i^2 = \sum_{i=1}^{M} [y_i - (ax_i + b)]^2$$

where $m(a,b)$ is the function of the parameters a and b that is to be minimized.

As in the calculus, we differentiate with respect to each of the parameters and set the derivatives equal to zero.

$$\frac{\partial m}{\partial a} = -2 \sum_{i=1}^{M} [y_i - (ax_i + b)] x_i = 0$$

$$\frac{\partial m}{\partial b} = -2 \sum_{i=1}^{M} [y_i - (ax_i + b)] = 0$$

Dropping the -2 factors and rearranging, we have

$$a \sum_{i=1}^{M} x_i^2 + b \sum_{i=1}^{M} x_i = \sum_{i=1}^{M} x_i y_i$$

$$a \sum_{i=1}^{M} x_i + b \sum_{i=1}^{M} 1 = \sum_{i=1}^{M} y_i$$

as the pair of linear equations to be solved for the unknown values of the parameters a and b.

NOTE: $\sum_{i=1}^{M} 1 = M$ and the symmetry of the coefficients.

These equations are called the "normal equations," and their solution determines the least-squares straight line (in this case).

x	y	x^2	xy
0	0	0	0
1	1	1	1
2	1	4	2
3	4	9	12
4	4	16	16
Sum 10	10	30	31

$$a \times 30 + b \times 10 = 31 \quad | \; +1$$
$$a \times 10 + b \times 5 = 10 \quad | \; -2$$
$$10a = 11$$

$$\boxed{a = \frac{11}{10} = 1.1}$$

$$5b = 10 - 10a$$
$$= -1$$
$$\boxed{b = -\tfrac{1}{5}}$$

$\therefore$ *the line is*

$$y = 1.1x - \tfrac{1}{5}$$

PROBLEM 10.2

1 Find the least-square straight line fitting the data

x	y
0	2
1	5
2	4
3	6
4	9

10.3 Polynomial approximation

A general polynomial of (fixed) degree N through $M(> N)$ points is easy to do because the parameters (the coefficients of the polynomial) occur linearly in the function, and as a result they also occur linearly in the **normal equations** that come from the differentiation process.

To see how the method goes for the general polynomial, let the equation of the polynomial be

$$y = a_0 + a_1 x + a_2 x^2 + \cdots + a_N x^N$$

and let the data be

$$(x_i, y_i) \qquad i = 1, \ldots, M$$

We wish to minimize

$$m(a_1, \ldots, a_N) \equiv m\ (a_j) = \sum_{i=1}^{M} \varepsilon_i^2$$

$$= \sum_{i=1}^{M} [y_i - (a_0 + a_1 x_i \cdots + a_N x_i^N)]^2$$

with respect to the parameters $a_1, \ldots, a_n$. Differentiating $m(a_j)$ with respect to each a_k and setting the result equal to zero, we have the $N + 1$ equations

$$\frac{\partial m(a_j)}{\partial a_k} = -2 \sum_{i=1}^{M} \left[y_i - \left(a_0 + a_1 x_i + \cdots + a_N x_i^N \right) \right] x_i^k$$

$$= 0 \qquad k = 0, 1, 2, \ldots, N$$

Rearranging and writing out in more detail, we have the corresponding normal equations

$$a_0 \sum 1 \ + a_1 \sum x_i \ + \cdots + a_N \sum x_i^N \ = \sum y_i$$

$$a_0 \sum x_i \ + a_1 \sum x_1^2 \ + \cdots + a_N \sum x_i^{N+1} = \sum y_i x_i$$

$$\cdots\cdots\cdots\cdots\cdots\cdots\cdots\cdots\cdots\cdots\cdots\cdots\cdots\cdots\cdots\cdots$$

$$a_0 \sum x_i^N + a_1 \sum x_i^{N+1} + \cdots + a_N \sum x_i^{2N} = \sum y_i x_i^N$$

To simplify the notation we set

$$\sum x_i^k = S_k \qquad (k = 0, 1, \ldots, 2N)$$

$$\sum x_i^k y_i = T_k \qquad (k = 0, 1, \ldots, N)$$

The $N + 1$ equations can now be written as

$$
\begin{aligned}
S_0 a_0 + S_1 a_1 &+ \cdots + S_N a_N &= T_0 \\
S_1 a_0 + S_2 a_1 &+ \cdots + S_{N+1} a &= T_1 \\
&\cdots\cdots\cdots\cdots\cdots\cdots\cdots\cdots \\
S_N a_0 + S_{N+1} a_1 &+ \cdots + S_{2N} a_N &= T_N
\end{aligned}
$$

We need to solve these equations for the $N + 1$ unknowns $a_0, a_1, \ldots, a_N$. Note that there are only $3N + 2$ sums to be found ($2N + 1$ S's and $N + 1$ T's).

It is easy to show that the determinant of these equations is not zero, for if the determinant were zero, then the homogeneous equations (with the right-hand sides all zero) would have a nonzero solution. This follows from the observations that $D = 0$ implies that any solution of the first $N - 1$ equations will automatically satisfy the last equation and that we can therefore assign any value we please to one of the variables in the first $N - 1$ equations.

To show that the homogeneous equations have only the zero solution, we multiply the first homogeneous equation by a_0, the second by $a_1, \ldots$, and the last by a_N and then add all these equations to get

$$
\sum_k \sum_j a_k a_j S_{k+j} = \sum_k \sum_j a_k a_j \sum_i x_i^{k+j}
$$

$$
= \sum_i \left(\sum_k a_k x_i^k \right) \left(\sum_j a_j x_i^j \right)
$$

$$
= \sum_i y(x_i) y(x_i)
$$

$$
= \sum_i y^2 (x_i) = 0
$$

This requires that the polynomial $y(x)$ of degree N be zero for the $M > N$ values, $x_1, \ldots, x_M$; hence $y(x) = 0$ for all x. Thus the homogeneous equations have only the zero solution, and therefore the original determinant was not zero—which is what we set out to prove.

Although the determinant is theoretically not zero, for $N > 10$ it is so small that it might as well be zero. As long as N is less than, say, 6, it is reasonable to solve the system of equations; around 10 or so, it is often difficult; and by $N = 20$, it is almost impossible. One solution to this trouble is given in the next chapter, on orthogonal functions.

PROBLEM   10.3

1 Sketch a flow diagram for computing the sums S_k and T_k.

10.4 Weighted least squares

It is a common experience that all the data points are not equally reliable. Typically, they tend to get less reliable as one approaches one (or both) ends of the range of measurements. One method of taking care of this matter of variable accuracy is to attach suitable weights $w_i \geq 0$ to each term in the sum

$$\sum_{i=1}^{M} w_i \varepsilon_i^2$$

that is being minimized. A brief examination of the process shows that the effect is merely that the sums

$$S_k = \sum_{i=1}^{M} w_i x_i^k$$

$$T_k = \sum_{i=1}^{M} w_i y_i x_i^k$$

now include the weights w_i. Otherwise, there are no significant changes.

PROBLEM 10.4

1 Fill in the details of the derivation in this section.

10.5 The general linear case

Consider, next, the more general case in which the unknown coefficients a_k occur linearly. In this case we are approximating (fitting, smoothing) with the function

$$y(x) = a_1 f_1(x) + a_2 f_2(x) + \cdots + a_N f_N(x)$$

where the $f_j(x)$ are given functions of x. If the given data are (x_i, y_i), $i = 1, \ldots, M$ and $M > N$, then we want to minimize

$$m(a_k) = \sum_{i=1}^{M} \varepsilon_i^2,$$

$$= \sum_{i=1}^{M} \left\{ y_i - [a_1 f_1(x_i) + a_2 f_2(x_i) + \cdots + a_N f_N(x_i)] \right\}^2$$

As usual we differentiate with respect to each of the unknowns a_k in turn and set the derivatives equal to zero (and neglect the factor of -2). We get

$$a_1 \sum f_1 f_j + a_2 \sum f_2 f_j + \cdots + a_N \sum f_N f_j = \sum y_i f_j$$

for $j = 1, \ldots, N$. The summations are over i, the index of the sample points. We set

$$S_{k,j} = \sum_{i=1}^{M} f_k(y_i) f_j(x_i) = S_{j,k}$$

$$T_j = \sum_{i=1}^{M} y_i f_j(x_i)$$

and obtain the "normal equations"

$$
\begin{aligned}
S_{1,1} a_1 + S_{1,2} a_2 + &\cdots + S_{1,N} a_N = T_1 \\
S_{2,1} a_1 + S_{2,2} a_2 + &\cdots + S_{2,N} a_N = T_2 \\
&\cdots \\
S_{N,1} a_1 + S_{N,2} a_2 + &\cdots + S_{N,N} a_N = T_N
\end{aligned}
$$

In the general case we can prove that the determinant is not zero in the same way as before **provided** the $f_j(x_i)$ are all linearly independent (over the set of points x_i).

10.6 Nonlinear parameters

If the parameters of the family of curves do not occur in a linear fashion, then the normal equations are no longer linear and are hence usually more difficult to solve. For these cases we can use the methods in Chap. 9.

The best simple strategy if there are some nonlinear parameters and some linear parameters seems to be as follows: Guess at a set of values for the nonlinear parameters, and then do a least-squares fit on the linear parameters. Regard the resulting sum of the squares of the residuals as a function of the nonlinear parameters **only**, and minimize it by the methods in Chap. 9.

Examples

$$y = a + be^{cx}$$

$$y = \frac{a}{1 + bx^2}$$

$$y = \frac{a_1}{1 + b_1 x^2} + \frac{a_2}{1 + b_2 x^2} + \frac{a_3}{1 + b_3 x^2}$$

In the case

$$y = a + be^{cx}$$

1 *Guess at a c value.*
2 *Fit a and b as before.*
3 *Plot the sum of squares corresponding to c.*
4 *Change c, using gradient*

$$\frac{\partial m}{\partial c} = \sum bxe^{cx} (y - a - be^{cx})$$

and repeat cycle from 2 on.

In this case the function and gradient evaluation includes the minimization of the linear parameters each time.

10.7 General remarks

It frequently happens that the system of equations does not determine the parameters very accurately. This tends to worry the beginner.

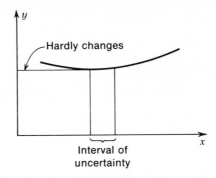

The reason they are poorly determined is often that the minimum is broad and rather flat. It is true that the optimum values are then poorly known, but it is also true that whichever set you choose (among those that look good) does not very much affect the value of the function you were minimizing; thus, the uncertainty does you little harm.

The method of least squares seems to be much more complex to the beginner than it is. The method is: You simply write down the expression for the sum of the squares of the residuals, decide on the parameters you have available, and then minimize the expression as a function of these parameters. The algebra may get messy, but the ideas are very simple.

$$m = \sum_{i=1}^{M} \varepsilon_i{}^2$$

$$= \sum_{i=1}^{M} [y_i - y\,(x_2)]^2$$

The student who does several least-squares fittings will soon notice that the biggest errors often occur at the ends of the interval, which can be annoying. Using suitable weights can cure this if necessary.

10.8   The continuous case

Sometimes we want to fit a function in an entire interval by a least-squares approximation. Consider the specific problem of fitting

$$y = e^x \qquad (0 \le x \le 1)$$

by the straight line

$$Y = a + bx$$

Proceeding as usual, we form the square of the residuals and "sum" over all the data; that is, we integrate over the range

$$m(a,b) = \int_0^1 [e^x - (a + bx)]^2 \, dx$$

Now we differentiate with respect to the linear parameters

$$\frac{\partial m}{\partial a} = -2 \int_0^1 [e^x - a - bx] \, 1 \, dx = 0$$

$$\frac{\partial m}{\partial b} = -2 \int_0^1 [e^x - a - bx] x \, dx = 0$$

Upon doing the integrations, we get (after dropping the factor -2)

$$e - 1 = a + \frac{b}{2}$$

$$1 = \frac{a}{2} + \frac{b}{3}$$

Solving these equations, we get

$$b = 6(3 - e) \approx 1.69$$

$$a = 4e - 10 \approx 0.873$$

Hence,

$$Y = 0.873 + 1.69x$$

is the least-squares approximating polynomial. The use of a positive (or at least a nonnegative) weight function in the continuous case follows exactly in the same process as in the discrete case.

PROBLEM 10.8

1 Approximate $y = e^x$, $0 \leq x \leq 1$, by a quadratic $y = a + bx + cx^2$ in the least-squares sense.

ORTHOGONAL FUNCTIONS

11.1 Introduction

The idea of a set of orthogonal functions plays a central role in both pure and applied mathematics, as well as in numerical computation. As we shall show in this chapter, orthogonal functions provide a powerful alternative to the direct approach of the least-squares approximation to a set of data. The idea of orthogonal functions is also used in the next two chapters. For these reasons it is necessary to devote much of this chapter to theoretical results rather than to the immediately practical.

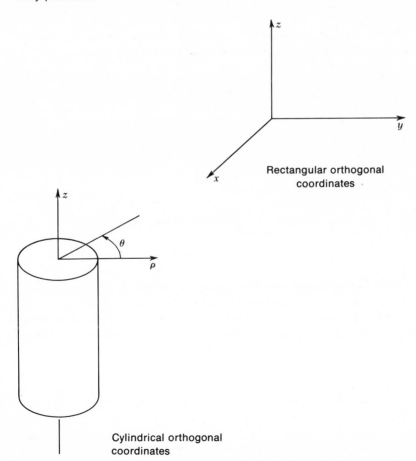

Rectangular orthogonal
coordinates

Cylindrical orthogonal
coordinates

11.2   The idea of orthogonal functions

Two functions† $f_1(x)$ and $f_2(x)$ are said to be orthogonal on an interval $a \leq x \leq b$ if

$$\int_a^b f_1(x) f_2(x) \, dx = 0$$

A set of functions $f_j(x)$, $(j = 0, 1, \ldots)$, is said to be orthogonal if the functions are mutually orthogonal, that is, if

$$\int_a^b f_j(x) f_k(x) \, dx = 0 \qquad (j \neq k)$$

Suppose we wish to represent a given function $F(x)$ in terms of a set of orthogonal functions $f_j(x)$ in the form of an expansion

$$F(x) = \sum_{j=0}^\infty a_j f_j(x)$$

To find the coefficients of the expansion, we multiply the equation by $f_k(x)$ and formally integrate‡ over the interval of orthogonality,

$$\int_a^b F(x) f_k(x) \, dx = \sum_{j=0}^\infty a_j \int_a^b f_j(x) f_k(x) \, dx$$

$$= a_k \int_a^b f_k^2(x) \, dx$$

since all the other terms vanish owing to the orthogonality of the $f_j(x)$. Hence, we have the coefficients

$$a_k = \frac{\displaystyle\int_a^b F(x) f_k(x) \, dx}{\displaystyle\int_a^b f_k^2(x) \, dx}$$

of the formal expansion. We are ignoring details of rigor.

This form of representation may be compared with the representation of a vector in three dimensions as a linear combination of mutually orthogonal unit vectors along the coordinate system, that is, in terms of the components of the vector.

---

†The function $f(x) \equiv 0$, $(a \leq x \leq b)$, is excluded from consideration as an orthogonal function since it is clearly orthogonal to all functions including itself.

---

‡We should sum in the discrete-data case.

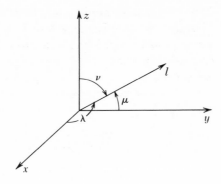

Two lines l_1 and l_2 are orthogonal if

$$\lambda_1 \lambda_2 + \mu_1 \mu_2 + \nu_1 \nu_2 < 0$$

With orthogonal functions we have an infinite dimensional space where there can be an infinite number of functions $f_j(x)$ in the set of functions. With either the vectors or the orthogonal function, we find the coefficients of the expansion by similar formulas, **dot products**, as they are called.

A vector **v** *can be represented as*

$$\mathbf{v} = v_x \mathbf{i} + v_y \mathbf{j} + v_z \mathbf{k}$$

A dot product of vectors **u** *and* **v** *can be represented by*

$$\mathbf{u} \cdot \mathbf{v} = u_x v_x + u_y v_y + u_z v_z$$

   The theory of orthogonal functions may be developed in terms of either a continuous interval $a \le x \le b$ or a discrete set of points x_i.

Either

$$\int_a^b f_k(x) f_j(x)\, dx = 0$$

or

$$\sum_{i=1}^{M} f_k(x_i) f_j(x_i) = 0$$

The theories are formally very much the same; notationally, the continuous interval is easier to follow, so we will use it. It is necessary to note, however, that with a discrete set of points there can be, at most, as many functions $f_j(x)$ in the set as there are points x_i in the discrete set of data.

It is sometimes convenient to include a weight function in the formulas

$$\int_a^b w(x)f_j(x)f_k(x)\ dx = 0 \qquad (j \ne k)$$

where

$$w(x) \ge 0 \qquad \text{in } (a \le x \le b)$$

just as we did in the least-squares theory. There is no essential difference between the general weight function and the special one

$$w(x) \equiv 1$$

so we will generally omit writing the weight factor.

$$a_k = \frac{\displaystyle\int_a^b w(x)F(x)f_k(x)\ dx}{\displaystyle\int_a^b w(x)f_k{}^2(x)\ dx}$$

11.3 An example, the Legendre polynomials

The Legendre polynomials, which are written as $P_j(x)$, are an important set of orthogonal functions over the interval $-1 \le x \le 1$. The first few of them are

$$P_0(x) = 1$$

$$P_1(x) = x$$

$$P_2(x) = \frac{3x^2 - 1}{2}$$

$$P_3(x) = \frac{5x^3 - 3x}{2}$$

$$P_4(x) = \frac{35x^4 - 30x^2 + 3}{8}$$

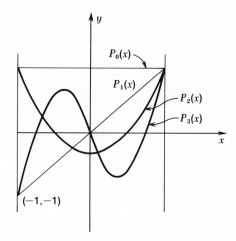

The general Legendre polynomial may be defined by Rodrigues' formula

$$P_n(x) = \frac{1}{2^n\,n!}\,\frac{d^n}{dx^n}\,(x^2 - 1)^n$$

which is clearly a polynomial of degree n.

Example

$$P_2(x) = \frac{1}{8}\,\frac{d^2}{dx^2}\,(x^2 - 1)^2$$

$$= \frac{1}{8}\,\frac{d^2}{dx^2}\,(x^4 - 2x^2 + 1)$$

$$= \frac{1}{8}(12x^2 - 4)$$

$$= \frac{3x^2 - 1}{2}$$

We need to show that they are orthogonal. Let $k < j$

$$I = \int_{-1}^{1} P_k(x)\,P_j(x)\,dx$$

$$= \frac{1}{2^k\,k!}\,\frac{1}{2^j\,j!}\int_{-1}^{1}\frac{d^k}{dx^k}\,(x^2 - 1)^k\,\frac{d^j}{dx^j}\,(x^2 - 1)^j\,dx$$

For convenience, set

$$\frac{1}{2^k k!} \frac{1}{2^j j!} = C$$

and integrate by parts, using

$$U = \frac{d^k}{dx^k} (x^2 - 1)^k \qquad dV = \frac{d^j}{dx^j} (x^2 - 1)^j \, dx$$

Some details

$$\int_{-1}^{1} \frac{d^k}{dx^k} (x^2 - 1)^k \frac{d^j}{dx^j} (x^2 - 1)^j \, dx = \frac{d^k}{dx^k} (x^2 - 1)^k \frac{d^{j-1}}{dx^{j-1}} (x^2 - 1)^j \Big|_{-1}^{1}$$

$$- \int \frac{d^{k+1}}{dx^{k+1}} (x^2 - 1)^k \frac{d^{j-1}}{dx^{j-1}} (x^2 - 1)^j \, dx$$

zero at both limits

Since the $(j - 1)$st derivative of $(x^2 - 1)^j$ had a factor of $x^2 - 1$, the integrated part drops out at both limits.

$$I = -C \int_{-1}^{1} \frac{d^{k+1}}{dx^{k+1}} (x^2 - 1)^k \frac{d^{j-1}}{dx^{j-1}} (x^2 - 1)^j \, dx$$

We can repeat the process j times until we get

$$I = C(-1)^j \int_{-1}^{1} \left[\frac{d^{k+j}}{dx^{k+j}} (x^2 - 1)^k \right] (x^2 - 1)^j \, dx$$

But since $j > k$, then

$$\frac{d^{k+j}}{dx^{k+j}} (x^2 - 1)^k = 0$$

and the functions are orthogonal over the inverval $-1 \le x \le 1$.
The case $j = k$ is needed to complete the matter. We have

$$I_k = \int_{-1}^{1} P_k{}^2(x) \, dx$$

$$= \left(\frac{1}{2^k k!} \right)^2 \int_{-1}^{1} \frac{d^k}{dx^k} (x^2 - 1)^k \frac{d^k}{dx^k} (x^2 - 1)^k \, dx$$

Integrating by parts as before, we have at the final differentiation

$$I_k = \frac{(2k)!}{(2^k k!)^2} \int_{-1}^{1} (1 - x^2)^k \, dx$$

where we have moved the $(-1)^k$ into the integrand. We now set $x = \cos \theta$ to get

$$I_k = \frac{(2k)!}{(2^k k!)^2} \int_{0}^{\pi} \sin^{2k+1} \theta \, d\theta$$

The integral is the well-known Wallis integral and has the value

$$2 \frac{2k(2k - 2)}{(2k + 1)(2k - 1)} \cdots \frac{2}{3(1)} = 2 \frac{(2^k k!)(2^k k!)}{(2k + 1)}$$

Thus, we finally have

$$I_k = \frac{2}{2k + 1}$$

From this it follows that when we make an expansion of an arbitrary function $F(x)$ in Legendre polynomials,

$$F(x) = \sum_{k=0}^{\infty} a_k P_k(x)$$

the coefficients are found by the formula

$$a_k = \frac{2k + 1}{2} \int_{-1}^{1} F(x) P_k(x) \, dx$$

Example   *Expand $F(x) = 1/(1 + x^2)$ in Legendre polynomials.*

$$\frac{1}{1 + x^2} = a_0 P_0 + a_1 P_1 + a_2 P_2 + \cdots$$

The coefficients are given by

$$a_k = \frac{2k + 1}{2} \int_{-1}^{1} \frac{P_k(x)}{1 + x^2} \, dx$$

For $k = 0$,

$$a_0 = \frac{1}{2} \int_{-1}^{1} \frac{dx}{1 + x^2}$$

$$= \frac{1}{2} \arctan x \Big|_{-1}^{1} = \frac{\pi}{2}$$

For k = 1,

$$a_1 = \frac{3}{2} \int_{-1}^{1} \frac{x}{1 + x^2} \, dx$$

$$= \frac{3}{2} \frac{\ln(1 + x^2)}{2} \bigg|_{-1}^{1}$$

$$= 0$$

Indeed the coefficient is zero for all odd k's. For

$$a_{2k} = \frac{4k + 1}{2} \int_{-1}^{1} \frac{x^{2k}}{x^2 + 1} \, dx$$

$$= \frac{4k + 1}{2} \int_{-1}^{1} x^{2k-2} \, dx - \frac{4k + 1}{2} \int_{-1}^{1} \frac{x^{2k-2}}{x^2 + 1} \, dx$$

It is not practical to compute the higher-order Legendre polynomials by the derivative formula; rather we can use the three-term recurrence relation which will be discussed in Sec. 11.8.

PROBLEMS 11.3

1 Expand e^x, $-1 \leq x \leq 1$, in the first three Legendre polynomials.
2 Compute $P_5(x)$ from the Rodrigues formula.

11.4 Linear independence

Linear independence is one of the most basic concepts in mathematics. A set of functions† $g_k(x)$ is said to be **linearly dependent** in an interval $a \leq x \leq b$ if there exists a set of constants c_k (not all zero) such that

$$\sum_{k=0}^{N} c_k g_k(x) \equiv 0 \qquad (a \leq x \leq b)$$

Otherwise the $g_k(x)$ are said to be **linearly independent**.

Any finite set of orthogonal functions $f_k(x)$ is automatically linearly independent. If any were not, then there would exist c_k (not all zero) such that

$$\sum_{k=0}^{N} c_k f_k(x) = 0$$

---

†Again it is assumed that $g_k(x) \not\equiv 0$.

But the c_k are given by

$$c_k = \frac{\int_a^b 0 f_k(x)\,dx}{\int_a^b f_k^2(x)\,dx} = 0$$

for every c_k.

The converse is also, in a sense, true: **Given a set of linearly independent functions** $g_k(x)$ **we can construct** (by the Gram-Schmidt process given below) **a set of orthogonal functions**. The construction is recursive. As a basis for the recursion, let

$$f_0(x) = g_0(x)$$
$$f_1(x) = g_1(x) - a_0^{(1)} f(x)$$

We require

$$\int_a^b f_0 f_1\,dx \equiv \int_a^b f_0(g_1 - a_0^{(1)} f_0)\,dx = 0$$

This means that

$$0 = \int_a^b f_0 g_1\,dx - a_0^{(1)}\int f_0^2\,dx$$

Hence,

$$a_0^{(1)} = \frac{\int_a^b f_0 g_1\,dx}{\int_a^b f_0^2\,dx}$$

For the general step in the recursion, suppose we have constructed $f_0, f_1, \ldots, f_k$ from $g_0, g_1, \ldots, g_k$ and want to find f_{k+1} by using g_{k+1}. Let

$$f_{k+1}(x) = g_{k+1}(x) - \sum_{j=0}^{k} a_j^{(k+1)} f_j(x)$$

where the $a_j^{(k+1)}$ are to be determined. We require for $m = 0, 1, \ldots, k$,

$$0 = \int_a^b f_{k+1} f_m\,dx$$

$$= \int_a^b g_{k+1} f_m\,dx - \sum_{j=0}^{k} a_j^{(k+1)}\int_a^b f_j f_m\,dx$$

Hence,

$$a_m^{(k+1)} = \frac{\displaystyle\int_a^b g_{k+1} f_m \, dx}{\displaystyle\int_a^b f_m{}^2 dx} \qquad (m = 0, 1, \ldots, k)$$

and we have found one more in the sequence of orthogonal functions. It is often convenient to scale a set of orthogonal functions so that

$$\int_a^b f_n{}^2 \, dx = 1$$

Such functions are called **orthonormal**.

To normalize a set $f_i(x)$, set

$$f_i{}^2(x) \, dx = \lambda_i > 0$$

Define the new set $\overline{f}_i(x)$ as

$$\overline{f}_i(x) = \frac{f_i(x)}{\sqrt{\lambda_i}}$$

Then

$$\int_a^b \overline{f}_i{}^2 \, dx = 1$$

The reason orthogonal functions play an important role in practical computing is that in a sense they are "more linearly independent" than the usual set of independent functions from which they are constructed. For example, in the interval $(-1 \leq x \leq 1)$, the functions x^n are similar to one another, but the Legendre polynomials differ rather more from each other than do the x^n.

PROBLEMS 11.4

1 Construct the first three Legendre polynomials by the orthogonalization process. Note that they differ from those listed by suitably chosen scale factors.

2 Modify the Gram-Schmidt process to construct orthonormal functions.

11.5   Least squares

Orthogonal functions are closely related to least squares, as is shown by the following theorem:

Theorem   *In an orthogonal function approximation*

$$F(x) = \sum_{j=0}^{\infty} a_j f_j(x)$$

the coefficients of the least-squares fit are given by the standard formula

$$a_j = \frac{\int_a^b F(x) f_j(x)\ dx}{\int_a^b f_j^2(x)\ dx}$$

Proof   As always in least squares we want to minimize

$$m = \int_a^b \left[F(x) - \sum_{j=0}^{\infty} a_j f_j(x) \right]^2 dx$$

We proceed in the standard way

$$\frac{\partial m}{\partial a_k} = -2 \int_a^b \left[F(x) - \sum_{j=0}^{\infty} a_j f_j(x) \right] f_k(x)\ dx = 0$$

or

$$\int_a^b F(x) f_k(x)\ dx = \sum_{j=0}^{\infty} a_j \int_a^b f_j(x) f_k(x)\ dx$$

$$= a_k \int_a^b f_k^2(x)\ dx$$

The first importance of this theorem is that whenever the direct process of least-squares fitting leads to a system of equations for the unknown coefficients that is hard to solve, then the equivalent orthogonal-function approach (which leads to equations that are trivial to solve) may be used. In one sense, the trouble has been pushed from solving the equations in the particular case to that of constructing the orthogonal functions in the general case.

The Gram-Schmidt process of constructing the orthogonal set often gives roundoff trouble, and experience seems to indicate that for polynomial systems the three-term recurrence relation discussed in Sec. 11.8 provides a preferable approach in constructing them. However, once having found the orthogonal functions appropriate to the interval (set of points), then any specific set of data is easily processed.

The second importance of this theorem is that we can determine the least-squares fit one term at a time rather than as in the direct method where each time we change the degree of the polynomial we have to recompute all the coefficients (though not all the sums involved).

An even better method, based on quasi-orthogonal functions, will be given in Sec. 11.10.

11.6   The Bessel inequality

The quality of the least-squares fit with the use of orthogonal functions can be found in two ways. We will use a discrete set of points rather than a continuous interval to illustrate this result. The first direct way is to compute the sum of the squares of the residuals at the points $i = 1, 2, \ldots, M$,

$$D^2 = \sum_{i=1}^{M} \varepsilon_i^2$$

$$= \sum_{i=1}^{M} \left[F(x_i) - \sum_{j=0}^{N} a_j f_j(x_i) \right]^2$$

In the second method we merely transform this expression by expanding it out and making some simple substitutions, using the notation

$$\lambda_j = \sum_{i=1}^{M} f_j^2 (x_i)$$

$$D^2 = \sum_{i=1}^{M} F^2(x_i) - 2 \sum_{j=0}^{M} a_j \sum_{i=1}^{M} F(x_i)f_j(x_i) + \sum_{j=0}^{N} \sum_{k=0}^{N} a_j a_k \sum_{i=1}^{M} f_j(x_i)f_k(x_i)$$

$$= \sum_{i=1}^{M} F^2(x_i) - 2 \sum_{j=0}^{N} a_j \lambda_j a_j + \sum_{j=0}^{N} a_j^2 \lambda_j$$

$$= \sum_{i=1}^{M} F^2(x_i) - \sum_{j=0}^{N} a_j^2 \lambda_j$$

Thus, if we first compute

$$\sum_{i=1}^{M} F^2(x_i)$$

by summing over all the data points and then subtract the square of each coefficient (times the corresponding λ_j), then the result is the sum of the squares of the residuals for that many terms in the approximation. If we now graph this quantity $D^2 = D^2(N)$, we often get a clue to whether or not to include more terms in the expansion.

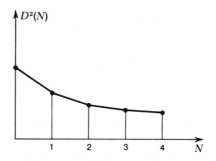

A better clue to when to stop in the process of fitting more and more terms in the least-squares approximations can be found from following the usual statistical practice of examining the residuals, in particular, examining the number of sign changes in the residuals.

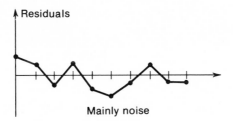

Mainly noise

Consider first the extreme condition of having only "noise" left in the residuals. Supposing that the residuals are random and independent of one another, we should, by using simple probability, expect that a given residual would be followed by one of the same sign half the time and by one of the opposite sign half the time. Thus we should expect to see about half the number of possible sign changes. If the number of sign changes were different from the number expected, either too low or too high, by several times the square root of the number of possible changes, then we should suspect that there was some kind of "signal" left and that our assumption of pure noise was wrong.

At the other extreme of no noise but all signal, then, at least for the typical orthogonal system of functions, we should expect that the residuals would have about one more sign change than had the last function we fitted, and a couple of extra sign changes above that would not worry us.

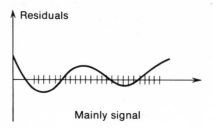

Mainly signal

The reason for this is that most orthogonal function systems have the property that the kth function has about k sign changes (this is exactly true of the important class of orthogonal polynomials; see Sec. 11.9).

What we shall see in the residuals, of course, is probably a mixture of the two extremes, both some signal and some noise.

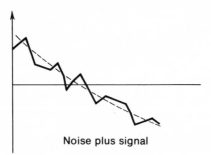

Noise plus signal

The noise oscillations will tend to increase the number of sign changes near the crossings due to the signal, and the more sign changes we see, the less we expect to reduce the sum of the squares if we fit a few more orthogonal functions.

This theory is a "down to earth" theory and is not given in the usual statistical courses. The test that is currently in statistical good graces is quite complex to understand, has some dubious assumptions (all tests must make some assumptions), and probably is of less use to the practicing scientist who is not highly trained in statistics and is therefore not in a position to appreciate what the orthodox test means. If you do understand advanced statistics, then by all means use it **if** it seems appropriate.

11.7 Orthogonal polynomials

If we choose the basic set of linearly independent functions as $y_k(x) = x^k$, $(k = 0, 1, \ldots)$ (the fundamental theorem of algebra shows that they are linearly independent), then the corresponding kth orthogonal function is a polynomial of degree k and the set is called an "orthogonal polynomial set." The Legendre polynomials form the orthogonal polynomial set for the interval $(-1 \le x \le 1)$ with weight $w(x) \equiv 1$.

It is easy to convert an arbitrary polynomial $y(x) = a_0 + a_1 x + \cdots + a_N x^N$ to a sum

$$y(x) = c_0 P_0 + c_1 P_1 + \cdots + c_N P_N$$

of orthogonal polynomials. Take the orthogonal polynomial of degree N, and subtract from $y(x)$ the appropriate multiple of the Nth orthogonal polynomial to make the term in x^N exactly vanish.

Example   Write x^3 in terms of Legendre polynomials, that is,

$$x^3 = a_0 P_0 + a_1 P_1 + a_2 P_2 + a_3 P_3$$

$$x^3 = x^3$$

$$-\tfrac{2}{5} P_3 = -x^3 + \tfrac{3}{5} x$$

$$\overline{\qquad\qquad \tfrac{3}{5} x \qquad}$$

$$-\tfrac{3}{5} P_1 = \qquad\quad -\tfrac{3}{5} x$$

$$\overline{\qquad\qquad 0 \qquad}$$

$$x^3 - \tfrac{2}{5} P_3 - \tfrac{3}{5} P_1 = 0$$

or

$$x^3 = \tfrac{2}{5} P_3 + \tfrac{3}{5} P_1$$

We then have the coefficient c_N of $P_N(x)$ in the expansion. [The term $P_N(x)$ no longer represents the Legendre polynomial system but now represents a typical orthogonal polynomial system.] The difference $y(x) - c_N P_N$ is a polynomial of degree $N - 1$, and we next remove the appropriate multiple of P_{N-1} to cause the term in x^{N-1} to vanish; and so forth.

Sometimes there are tables giving the powers of x as sums of orthogonal polynomials; that is,

$$x^k = a_0^{(k)} P_1 + a_1^{(k)} P_1 + \cdots + a_k^{(k)} P_k \qquad \text{for } k = 0, 1, \ldots$$

to make the conversion easy.

The reverse process of going from the expansion in the orthogonal polynomials to a single polynomial is merely a matter of substituting the polynomial representation of $P_k(x)$ in place of $P_k(x)$ and gathering together like terms in x. The processes are simple in theory; in practice the loss of accuracy due to cancellation of large numbers is another matter.

It should be clear now that the problem of least-squares fitting a polynomial to some data is equivalent to first finding the appropriate orthogonal polynomials and then expanding the data in the formal orthogonal polynomial representation.

Since this is so important a process, we will investigate the class of orthogonal polynomials a bit more. We shall prove that[†] the *nth polynomial $f_n(x)$ has exactly n real, distinct zeros in the interval (for the discrete set of points, n changes in sign).* To prove this, assume the opposite, that there are only $k < n$ distinct zeros. Corresponding to each zero of odd multiplicity, include one factor in

$$q_k(x) = (x - x_1)(x - x_2) \cdots (x - x_{k'})$$
$$= x^{k'} + \cdots$$

where k' is the number of odd multiplicity zeros and may be less than k.

Now consider

$$\int_a^b f_n(x) q_k(x) \, dx > 0$$

By hypothesis this has an integrand which does not change sign,[‡] and hence the integral is clearly positive. On the other hand, we can write

$$q_k(x) = \sum_{j=0}^{k} a_j f_j(x)$$

[†]It is necessary to have the weight function $w(x) \geq 0$, $a \leq x \leq b$.

[‡]We assumed $w(x) \geq 0$.

Hence, since $k < n$,

$$\int_a^b f_n q_k \, dx = \sum_{j=0}^k a_j \int_a^b f_n f_j \, dx = 0$$

which is a contradiction. Thus, $f_n(x)$ must have n real, distinct zeros in $a \le x \le b$.

11.8   The three-term recurrence relation

It is an important property of orthogonal polynomials (and often other orthogonal systems) that there is a simple linear relationship among any three consecutive polynomials. To prove this, consider the fact that the polynomial

$$xP_n(x)$$

has degree $n + 1$ and hence can be written as a sum of the first $n + 1$ orthogonal polynomials. We have

$$xP_n(x) = \sum_{k=0}^{n+1} a_k^{(n)} P_k(x)$$

To find the coefficient $a_k^{(n)}$ in the expansion, we have, as usual,

$$a_k^{(n)} = \frac{\displaystyle\int_a^b [xP_n(x)]P_k(x) \, dx}{\displaystyle\int_a^b P_k^2(x) \, dx}$$

But writing the numerator as

$$\int_a^b P_n(x) [xP_k(x)] \, dx$$

we see that for $k < n - 1$, $xP_k(x)$ is a sum of orthogonal polynomials of degree, at most, $k + 1$, which is less than n, and hence each is orthogonal to $P_n(x)$ and hence $a_k^{(n)} = 0$. We have, therefore, the desired proof that there is a three-term recurrence relationship among any three consecutive polynomials of the set; that is,

$$xP_n(x) = a_{n+1}^{(n)} P_{n+1}(x) + a_n^{(n)} P_n(x) + a_{n-1}^{(n)} P_{n-1}(x)$$

To obtain the actual coefficients of this expansion, suppose that the leading coefficient of the kth orthogonal polynomial is positive and is labeled b_k.

We note that (see Sec. 11.7)

$$xP_n(x) = \frac{b_n}{b_{n+1}} P_{n+1}(x) + \text{(lower-order polynomials)}$$

hence,

$$\int_a^b xP_n(x)P_{n+1}(x)\,dx = \frac{b_n}{b_{n+1}} \int_a^b P_{n+1}^2(x)\,dx$$

(the others being zero because of orthogonality). Hence†,

$$a_{n+1}^{(n)} = \frac{b_n \lambda_{n+1}}{b_{n+1} \lambda_n} > 0$$

For $a_{n-1}^{(n)}$ we proceed in much the same way

$$xP_{n-1} = \frac{b_{n-1}}{b_n} P_n(x) + \cdots$$

and conclude that

$$a_{n-1}^{(n)} = \frac{b_{n-1}}{b_n} \frac{\displaystyle\int_a^b P_n{}^2(x)\,dx}{\displaystyle\int_a^b P_{n-1}^2(x)\,dx}$$

Now let c_k be the coefficient of x^{k-1} in $P_k(x)$. Since the three-term relation is an identity, we may equate the coefficients of x^n in

$$xP_n = a_{n+1}^{(n)} P_{n+1} + a_n^{(n)} P_n + a_{n-1}^{(n)} P_{n-1}$$

to get

$$c_n = a_{n+1}^{(n)} c_{n+1} + a_n^{(n)} b_n$$

---

†We assumed $b_n > 0$ (all n).

Now, using the value we just found of $a_{n+1}^{(n)} = (b_n \lambda_{n+1})/(b_{n+1} \lambda_n)$ and solving for $a_n^{(n)}$ we get

$$a_n^{(n)} = \frac{c_n}{b_n} - \frac{c_{n+1}}{b_{n+1}} \frac{\lambda_{n+1}}{\lambda_n}$$

Thus, we have the coefficients of the three-term relation explicitly in terms of the leading coefficients of the polynomials.

11.9 Interlacing zeros

From the three-term recurrence relationship, we now wish to show that the zeros of $P_n(x)$ **lie between** the zeros of $P_{n+1}(x)$.

As a basis for induction, we observe that

$$P_0(x) = b_0 > 0$$

has no zeros and

$$P_1(x) = b_1 x + \text{constant}$$

has one real zero in the range (a,b), otherwise, $\int P_0 P_1 \, dx \neq 0$. Suppose the result is true for $P_n(x)$. At the n zeros of $P_n(x)$, the three-term relation is

$$a_{n+1} P_{n+1}(x) + a_{n-1} P_{n-1}(x) = 0$$

or

$$P_{n+1}(x) = -\frac{a_{n-1}}{a_{n+1}} P_{n-1}(x)$$

$$= -\frac{b_{n-1} b_{n+1} \lambda_n^2}{b_n^2 \lambda_{n-1} \lambda_{n+1}} P_{n-1}(x)$$

where b_k is the leading coefficient of $P_k(x)$. The coefficient of $P_{n-1}(x)$ is negative. Thus, $P_{n-1}(x)$ and $P_{n+1}(x)$ have opposite signs at the zeros of $P_n(x)$. Starting now at $x = b$ and working back, by the induction hypotheses we have found that $P_{n-1}(x)$ is still positive at the last zero of $P_n(x)$; hence, $P_{n+1}(x)$ is negative and has already passed through one zero. Each time we pass to the next lower zero of $P_n(x)$, we know, by the induction hypotheses, that $P_{n-1}(x)$ has had another zero and therefore has changed sign. Hence $P_{n+1}(x)$ must also have had a zero. And so we go on; between each pair of zeros of $P_n(x)$ is a zero of $P_{n+1}(x)$; that is, the zeros of $P_n(x)$ are between the zeros of $P_{n+1}(x)$.

11.10 Quasi-orthogonal polynomials

The advantages of using orthogonal polynomials in the least-squares fitting of a polynomial to some data is now obvious. The disadvantage that one must construct the appropriate set of orthogonal functions for the set of sample points used is also obvious. For equally spaced data, extensive tables have been published.[†] However, often the data is irregular, and one wonders how to avoid both troubles.

Some consideration of the two situations—that the direct approach leads to a system of simultaneous linear equations which is apt to be hard to solve in practice, whereas the orthogonal polynomial approach produces a diagonal system of equations which is trivial to solve—suggests that what is wanted is some easy method of getting an easily solved system. How can we make the off-diagonal terms of the system of normal equations very small while the diagonal terms are fair sized?

One approach is to construct a family of quasi-orthogonal polynomials by a sequence of polynomials having interlacing zeros. Thus in the interval, say, $-1 \leq x \leq 1$, picking the quasi-orthogonal polynomials

$$p_0 = 1$$
$$p_1 = x$$
$$p_2 = (x - \tfrac{1}{2})(x + \tfrac{1}{2})$$
$$p_3 = (x - \tfrac{2}{3})(x + \tfrac{2}{3})x$$
etc.

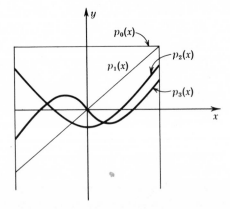

will lead to a system of easily solved equations. On the other hand, since we made up the polynomials and can choose as convenient coefficients as we please, we can arrange to have relatively easy arith-

---

†R. A. Fisher, and Yates,

metic. Thus, we have a pleasant avoidance of the worst troubles of each method.

Notice, for example,

$$\int_{-1}^{1} p_3(x)\, p_1(x)\; dx = \int_{-1}^{1} \left(x^4 - \tfrac{4}{9} x^2\right)\, dx$$

$$= \frac{x^5}{5} - \frac{4x^3}{27}\; \Big|_{-1}^{1}$$

$$= \frac{14}{135}$$

In general matrix is

$$\begin{pmatrix} 2 & 0 & \frac{1}{6} & 0 \\ 0 & \frac{2}{3} & 0 & \frac{14}{135} \\ \frac{1}{6} & 0 & \frac{7}{120} & 0 \\ 0 & \frac{14}{135} & 0 & \frac{322}{5502} \end{pmatrix}$$

The conversion from the quasi-orthogonal system to the polynomial form is accomplished by the definition of the polynomials. This method is little known but extremely useful and practical.

PROBLEM 11.10

1 Expand e^x in the quasi-orthogonal polynomials, and compare with Prob. 1, Sec. 11.3.

12.1 The continuous fourier series

Perhaps the most important set of orthogonal functions is

$$1, \cos x, \cos 2x, \cos 3x, \ldots$$
$$\sin x, \sin 2x, \sin 3x, \ldots$$

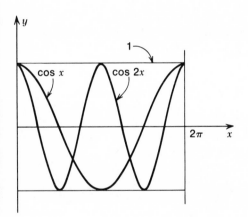

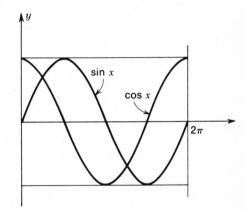

which leads to the Fourier series where the interval of orthogonality is $(0 \leq x \leq 2\pi)$ (although clearly because of the periodicity of the functions **any** interval of length 2π may be used). If, instead of $(0 \leq x \leq 2\pi)$, we use the interval $(0 \leq x \leq L)$, then we have the Fourier series

$$F(x) = \frac{a_0}{2} + a_1 \cos \frac{2\pi}{L}x + b_1 \sin \frac{2\pi}{L}x$$
$$+ a_2 \cos \frac{2\pi}{L}2x + b_2 \sin \frac{2\pi}{L}2x + \cdots$$

In this expansion the a_k and b_k are given by

$$a_k = \frac{2}{L} \int_0^L F(x) \cos \frac{2\pi}{L}kx \, dx \qquad (k = 0, 1, 2, \ldots)$$

$$b_k = \frac{2}{L} \int_0^L F(x) \sin \frac{2\pi}{L}kx \, dx \qquad (k = 1, 2, \ldots)$$

which are based on the orthogonality relations (m and n are, of course, integers)

$$\int_0^L \cos\frac{2\pi}{L}mx \, \cos\frac{2\pi}{L}nx \, dx = \begin{cases} 0 & m \neq n \\ \dfrac{L}{2} & m = n \neq 0 \\ L & m = n = 0 \end{cases}$$

$$\int_0^L \cos\frac{2\pi}{L}mx \, \sin\frac{2\pi}{L}nx \, dx = 0$$

$$\int_0^L \sin\frac{2\pi}{L}mx \, \sin\frac{2\pi}{L}nx \, dx = \begin{cases} 0 & m \neq 0 \\ \dfrac{L}{2} & m = n \neq 0 \\ 0 & m = n = 0 \end{cases}$$

These orthogonality relations are easily proved by using the trigonometric identity

$$\cos\frac{2\pi}{L}mx \, \cos\frac{2\pi}{L}nx = \frac{1}{2}\left[\cos\frac{2\pi}{L}(m+n)x + \cos\frac{2\pi}{L}(m-n)x\right]$$

and other related indentities for the various integrals. Carrying out the details of the first relations, we have for $m \neq n$,

$$\int_0^L \cos\frac{2\pi}{L}mx \, \cos\frac{2\pi}{L}nx \, dx = \frac{1}{2}\left[\frac{\sin(2\pi/L)(m+n)x}{(2\pi/L)(m+n)} + \frac{\sin(2\pi/L)(m-n)x}{(2\pi/L)(m-n)}\right]_0^L$$

$$= \begin{cases} 0 & m \neq n \\ \dfrac{L}{2} & m = n \neq 0 \\ L & m = n = 0 \end{cases}$$

The case of sin *times* cos

$$\int_0^L \sin\frac{2\pi}{L}mx \, \cos\frac{2\pi}{L}mx \, dx = \frac{1}{2}\int_0^L \left[\sin\frac{2\pi}{L}(m+n)x + \sin\frac{2\pi}{L}(m-n)x\right] dx$$

The integral $=0$ *from the periodicity and shape of* sin *functions.*

It is well known that any "resonable engineering function" can be expanded in a Fourier series.

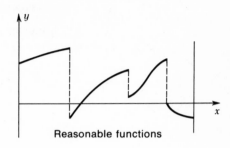

Reasonable functions

The rate of convergence of the series depends on the discontinuities in the function (and in its derivatives); but we must remember that the function is assumed to be periodic and is, as it were, wrapped around a cylinder.

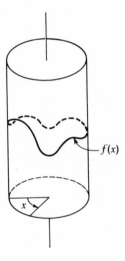

$f(x)$

Any interval of length L may be used; in particular, it is often useful to use $(-L/2 \le x \le L/2)$ instead of $(0 \le x \le L)$.

PROBLEMS 12.1

Find the Fourier expansions for:

1 $y = x, \quad (-1 \le x \le 1)$

2 $y = 1 - x^2, \quad (-1 \le x \le 1)$

3 $y = e^{-x}, \quad (0 \le x \le L)$

4 $y = \begin{cases} 1, & (0 < x < \pi) \\ 0, & (\pi < x < 2\pi) \end{cases}$

12.2   The complex form of the Fourier series

It is often convenient, especially for theory, to write the sin x and cos x in their equivalent complex forms.

Euler identity

$$e^{ix} = \cos x + i \sin x$$

Replace i with $-i$,

$$e^{-ix} = \cos x - i \sin x$$

Add and subtract these to get cos x and sin x.

$$\sin x = \frac{e^{ix} - e^{-ix}}{2i}$$

$$\cos x = \frac{e^{ix} + e^{-ix}}{2}$$

In this notation we have

$$F(x) = \sum_{k=-\infty}^{\infty} c_k e^{(2\pi i/L)kx} \, dx$$

where the c_k are found by the simple formula

$$c_k = \frac{1}{L} \int_0^L F(x) e^{-(2\pi i/L)kx} \, dx$$

The orthogonality condition which determines the form of the c_k is the easily proved

$$\int_0^L e^{(2\pi i/L)kx} e^{(2\pi i/L)mx} \, dx = \begin{cases} 0 & (m+k \neq 0) \\ L & (m+k = 0) \end{cases}$$

Comparing the two forms of representation, we find

$$c_k = \begin{cases} \dfrac{a_k - ib_k}{2} & (k > 0) \\[2mm] \dfrac{a_k + ib_k}{2} & (k < 0) \\[2mm] \dfrac{a_0}{2} & (k = 0) \end{cases}$$

PROBLEM   12.2

Expand in the complex Fourier series form:

1 $y = x$ $(\frac{1}{2}\pi < x < \pi)$

2 $y = \begin{cases} 1 & (0 \le x < \frac{1}{2}) \\ 0 & (\frac{1}{2} < x < 1) \end{cases}$

12.3   The finite Fourier series

It is a remarkable fact that the sines and cosines are orthogonal over both the continuous interval and **any** set of equally spaced discrete points.

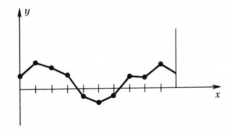

This is extremely important because in computing we are usually given only samples of the function on a set of equally spaced points. This means, in turn, that we can find the coefficients exactly (except for roundoff) in the discrete sampled problem, instead of approximately in the continuous model by numerical integration.

We will consider only an even number of points, $2N$.

Let the $2N$ sample points be

$$0, \frac{L}{2N}, \frac{2L}{2N}, \ldots, \frac{(2N-1)L}{2N,}$$

or more shortly,

$$x_p = \frac{Lp}{2N} \qquad (p = 0, 1, \ldots, 2N - 1)$$

We need to show that the $2N$ Fourier functions

$$1, \cos\frac{2\pi}{L}x, \cos\frac{2\pi}{L}2x, \ldots, \cos\frac{2\pi}{L}(N-1)x, \cos\frac{2\pi}{L}Nx$$

$$\sin\frac{2\pi}{L}x, \sin\frac{2\pi}{L}2x, \ldots, \sin\frac{2\pi}{L}(N-1)x$$

form an orthogonal set of functions. This amounts to showing that for $0 \le k, m \le N$

$$\sum_{p=0}^{2N-1} \cos\left(\frac{2\pi}{L}k\,\frac{Lp}{2N}\right) \cos\left(\frac{2\pi}{L}m\,\frac{Lp}{2N}\right) = \begin{cases} 0 & (k \ne m) \\ N & (k = m \ne 0, N) \\ 2N & (k = m = 0, N) \end{cases}$$

$$\sum_{p=0}^{2N-1} \cos\left(\frac{2\pi}{L}k\,\frac{Lp}{2N}\right) \sin\left(\frac{2\pi}{L}m\,\frac{Lp}{2N}\right) = 0$$

$$\sum_{p=0}^{2N-1} \sin\left(\frac{2\pi}{L}k\,\frac{Lp}{2N}\right) \sin\left(\frac{2\pi}{L}m\,\frac{Lp}{2N}\right) = \begin{cases} 0 & (k \ne m) \\ N & (k = m \ne 0, N) \\ 0 & (k = m = 0, N) \end{cases}$$

In fact we shall for future use prove more, namely, that when

$$k \pm m = 0, \pm 2N, \pm 4N, \ldots$$

the orthogonality conditions have special features.
    We begin simply by examining the series

$$\sum_{p=0}^{2n-1} e^{(2\pi i/L)jx_p} = \sum_{p=0}^{2N-1} e^{\pi i\,jp/N} \qquad \text{for integer } j$$

which is a geometric progression with ratio

$$r = e^{\pi i\,j/N}$$

whose sum is

$$\begin{cases} \dfrac{1 - r^{2N}}{1 - r} = 0 & (r \ne 1) \\[2mm] 2N & (r = 1) \end{cases}$$

This   is   true   because $e^{2\pi i} = 1$.   The   condition $r = 1$ means $j = 0, \pm 2N, \pm 4N, \ldots$.
    We next show that the set of functions

$$e^{(2\pi i/L)kx_p}$$

are orthogonal in the sense that the product of one function times the complex conjugate of another function summed over the points x_p is zero, that is, that

$$\sum_{p=0}^{2N-1} e^{(2\pi i/L)kx_p}\, e^{-(2\pi i/L)mx_p} = \begin{cases} 0 & (|k-m| \neq 0, 2N, 4N, \ldots) \\ 2N & (|k-m| = 0, 2N, 4N, \ldots) \end{cases}$$

This follows immediately from the above by writing the product as

$$\sum_{p=0}^{2N-1} e^{(2\pi i/L)(k-m)x_p}$$

and noting that $k - m$ plays the role of j.

We now return to the "real functions" by using the Euler identity

$$e^{ix} = \cos x + i\,\sin x$$

The condition for the single exponential summed over the points x_p becomes two equations (the real and imaginary parts separately)

$$\sum_{p=0}^{2N-1} \cos \frac{2\pi}{L} jx_p = \begin{cases} 0 & j \neq 0, \pm 2N, \pm 4N, \ldots) \\ 2N & (j = 0, \pm 2N, \pm 4N, \ldots) \end{cases}$$

$$\sum_{p=0}^{2N-1} \sin \frac{2\pi}{L} jx_p = 0 \qquad \text{for all } j$$

At last we are ready to prove the orthogonality of the Fourier functions over the set of equally spaced points x_p. The first of the three orthogonality equations can be written, by using the trigonometric identity, as

$$\cos a \cos b = \tfrac{1}{2}[\cos (a+b) + \cos (a-b)]$$

$$\tfrac{1}{2}\sum_{p=0}^{2n-1} \left[\cos \pi (k+m) \frac{p}{N} + \cos \pi (k-m) \frac{p}{N} \right]$$

$$= \begin{cases} 0 & (|k-m| \text{ and } |k+m| \neq 0, 2N, 4N, \ldots) \\ N & (|k-m| \text{ or } |k+m| = 0, 2N, 4N, \ldots) \\ 2N & (|k-m| \text{ and } |k+m| = 0, 2N, 4N, \ldots) \end{cases}$$

For the restricted set of functions we are now using, namely,

$$1, \cos \frac{2\pi}{L}x, \ldots, \cos \frac{2\pi}{L}(N-1)x, \cos \frac{2\pi}{L}Nx$$

$$\sin \frac{2\pi}{L}x, \ldots, \sin \frac{2\pi}{L}(N-1)x$$

$|k + m| = 0$, 2N, 4N, ... cannot occur unless $k = m = 0$ or N. Thus, we have the required orthogonality of the Fourier functions over the set x_p.

The orthogonality in turn leads directly to the expansion of an arbitrary function $F(x)$ defined on the set of points x_p,

$$F(x) = \frac{A_0}{2} + \sum_{k=1}^{N-1} \left(A_k \cos \frac{2\pi}{L} kx + B_k \sin \frac{2\pi}{L} kx \right) + \frac{A_N}{2} \cos \frac{2\pi}{L} Nx$$

where

$$A_k = \frac{1}{N} \sum_{p=0}^{2N-1} F(x_p) \cos \frac{2\pi}{L} kx_p \qquad (k = 0, 1, \ldots, N)$$

$$B_k = \frac{1}{N} \sum_{p=0}^{2N-1} F(x_p) \sin \frac{2\pi}{L} kx_p \qquad (k = 1, 2, \ldots, N-1)$$

The function $F(x)$ is often given at $2N + 1$ points, but still for $2N$ intervals, and it is assumed that

$$F(0) = F(L)$$

If this is not so, then it is customary to use

$$\frac{F(0) + F(L)}{2}$$

as the value of $F(0)$. In this case the formulas for the coefficients A_k are

$$A_k = \frac{1}{N} \left[\frac{F(0)}{2} + \sum_{p=1}^{2N-1} F(x_p) \cos \frac{2\pi}{L} kx_p + \frac{F(L)}{2} \right]$$

$$B_k = \frac{1}{N} \sum_{p=1}^{2N-1} F(x_p) \sin \frac{2\pi}{L} kx_p$$

These are effectively the trapazoid rule for numerically integrating the corresponding integrals in the continuous interval.

There is a second set of equally spaced points that we need to mention, namely,

$$t_p = x_p + \frac{L}{4N} = \frac{L}{2N} \left(p + \frac{1}{2} \right)$$

which are midway between the x_p points (recall that $x_0 = 0$ is the same as x_{2N} due to periodicity).

Repeating the steps briefly, we see that the sum

$$\sum_{p=0}^{2n-1} e^{(2\pi ij/L)\ell_p} = \begin{cases} 0 & (r \neq 1) \\ 2N & (r = 1) \end{cases}$$

as before. Hence, the rest follows as before except the special treatment of the end values in the summations for the coefficients.

PROBLEMS 12.3

Find the finite Fourier expansion for:

1   The data $x(0) = 0$, $x(1) = 1$, $x(2) = 2$, $x(3) = 3$; $N = 2$.
2   The data $x(\frac{1}{2}) = 1$, $x(\frac{3}{2}) = 1$, $x(\frac{5}{2}) = -1$, $x(\frac{7}{2}) = -1$; $N = 2$.

12.4   Relation of the discrete and continuous expansions

It is reasonable to ask: What is the relation between the two expansions we have found, the continuous and the discrete? Let the continuous expansion be

$$F(x) = \frac{a_0}{2} + \sum_{k=1}^{\infty} \left(a_k \cos \frac{2\pi}{L} x + b_k \sin \frac{2\pi}{L} x \right)$$

with lower-case letters for the coefficients. We pick $x_p = Lp/2N$ for convenience. If we multiply $F(x_p)$ by $\cos (2\pi/L)kx_p$ and sum, we get

$$\sum_{p=0}^{2N-1} F(x_p) \cos \frac{2\pi}{L} kx_p = NA_k = N(a_k + a_{2N-k} + a_{2N+k} \cdots)$$

Hence, the coefficient we calculate is

$$A_k = a_k + \sum_{m=1}^{\infty} (a_{2Nm-k} + a_{2Nm+k})$$

which expresses the (upper-case) finite Fourier series coefficients in terms of the (lower-case) continuous Fourier series coefficients.

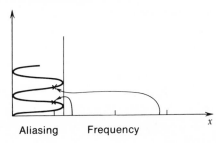

Aliasing   Frequency

Similarly,

$$B_k = b_k + \sum_{m=1}^{\infty} (-b_{2Nm-k} + b_{2Nm+k})$$

For the special constant term A_0, we have

$$A_0 = a_0 + 2 \sum_{m=1}^{\infty} a_{2Nm}$$

Thus, various frequencies present in the original continuous signal $F(x)$ are added together due to the sampling. This effect is called "aliasing" and *is directly attributable to the act of sampling at equally spaced points;* once the sampling has been done, the effect cannot be undone (from the samples alone).

This aliasing is a well-known effect to the watchers of TV westerns. The effect of the TV sampling is to make the stagecoach wheels appear to go backward at certain forward speeds. Rotating helicopter blades also show this effect on TV or in the movies. Aliasing is also the basis of the stroboscopic effect that is used to observe rapidly rotating machinery with a flashing light, which causes the machinery to appear as if it were rotating slowly.

12.5   The fast Fourier series

How shall we compute the Fourier coefficients? The direct approach seems to require about

$$(2N)^2$$

multiplications and additions. Recently a method has been popularized by J. W. Tukey and J. W. Cooley that significantly reduces the number of arithmetic operations needed to compute the finite Fourier coefficients (and because of the symmetry between the formulas for coefficients and those for the value of the series, it also applies to evaluating Fourier series for equally spaced points). If the number of sample points is a power of 2 (or has many small factors), then the number of arithmetic operations by this method can be reduced from approximately N^2 to around $N \log N$ operations. This can be very significant for long runs of data and has produced a fundamental change in what is currently practical.

In order to present the essential idea of the Tukey-Cooley algorithm, we shall simplify the notation we are using. We will suppose that the $x_p = Lp/2N$ and the length of the interval $L = 1$. We also assume that the number of sample points $2N$ (which could have been an odd number, but we assumed was even for convenience) can be factored,

$$2N = GH$$

Then our sample points are

$$x_p = \frac{p}{GH}$$

The coefficients of the Fourier expansion are

$$A_k = A(k) = \frac{1}{GH} \sum_{p=0}^{GH-1} F(x_p) e^{-2\pi i k x_p}$$

We now write

$$k = k_0 + k_1 G$$
$$p = p_0 + p_1 H$$

where $k_0 < G$, $k_1 < H$, $p_0 < H$, and $p_1 < G$. Then,

$$A(k_0 + k_1 G) = \frac{1}{GH} \sum_{p_0=0}^{H-1} \sum_{p_1=0}^{G-1} F\left(\frac{p_0 + p_1 H}{GH} \right) e^{-2\pi i (k_0 + k_1 G)(p_0 + p_1 H)/GH}$$

$$= \frac{1}{GH} \sum_{p_0=0}^{H-1} e^{-2\pi i k_0 p_0/GH} \, e^{-2\pi i k_1 p_0/H}$$

$$\times \left[\sum_{p_1=0}^{G-1} F\left(\frac{p_0}{GH} + \frac{p_1}{G} \right) e^{-2\pi i k_0 p_1/G} \right]$$

where we have used

$$e^{-2\pi i k_1 p_1} \equiv 1$$

We recognize that the contents of the brackets is the Fourier series of $1/H$ of the samples, phase-shifted p_0/GH. There are H such sums to be done.

If we label these sums as

$$\overline{A}(k_0, p_0)$$

we have

$$A(k_0 + k_1 G) = \frac{1}{GH} \sum_{p_0=0}^{H-1} \overline{A}(k_0, p_0) e^{-2\pi i \left[(k_0/GH) + (k_1/H) \right] p_0}$$

which is a second Fourier series, this time of H terms to be evaluated.

When we count the operations, we find them proportional to

$$GH(G + H)$$

Evidently, if the number of sample points had factors G_1, $G_2, \ldots, G_k$, we should have, by repeating this process,

$$G_1 G_2 \cdots G_k (G_1 + G_2 + G_3 + \cdots + G_k)$$

operations. When the number of sample points is a power of 2, we get the log factor that we earlier announced.

Much work has been done to optimize the fast Fourier series. Tukey and Cooley recommend finding factors of 2. Later examination of the problem of optimization produced the idea that it is better, when possible, to go by factors of 4; still later results suggest factors of 8 when possible. It is not the purpose of this book to try to produce the optimum library routine (because much still depends on the particular machine available) but rather to indicate reasonably efficient methods.

The details of a very efficient Tukey-Cooley method are given in *Mathematics of Computation*, volume 19, 1965, pages 297 to 307.

12.6   Cosine expansion

If we use the interval $(-L/2 \leq x \leq L/2)$, and from the definition of the function $(0 \leq x \leq L/2)$, define $F(x)$ to be an even function, that is,

$$F(-x) \equiv F(x)$$

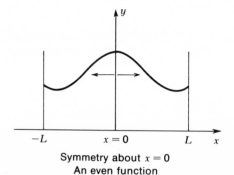

Symmetry about $x = 0$
An even function

(note that this does not introduce a discontinuity in the function), then the function is even and all sin terms will have zero coefficients.

Two   cases   occur   of   importance,   namely, $x_p = Lp/2N$ and $t_p = L(p + \frac{1}{2})/2N$.

For x_p,

$$A_k = \frac{1}{2N} \left[F(0) + 2 \sum_{p=0}^{N-1} F\left(\frac{pl}{2N}\right) e^{(-2\pi ik/2N)p} + F(L) \right]$$

In the case of t_p,

$$A_k = \frac{1}{N} \sum_{p=0}^{N-1} F\left(\frac{p + \frac{1}{2}}{2N}\right) e^{(-2\pi i/2N)[p + 1/2]}$$

The first is essentially the trapezoid rule, whereas the second is the midpoint formula. We shall use these in the next chapter.

CHEBYSHEV APPROXIMATION 13

13.1 Introduction

The words "Chebyshev approximation" are used when the approximating function is chosen so that the maximum error is minimized, rather than, as in least-squares approximation, so that the sum of the squares of the errors is minimized. It is sometimes called the "minimax approximation" because it minimizes the maximum error.

Chebyshev

Minimize {max error}

Least squares

Minimize {sum of squares of errors}

The idea of Chebyshev approximation is very popular in many areas of work. Thus, game theory selects the strategy (or strategies) that minimizes the maximum loss. In the military this same minimum-loss criterion seems to be widely accepted for planning. In computing, especially in basic function-evaluation routines for $\sin x$, e^x, $\log x$, and so forth, the user wants to be assured that his error is never more than a given amount: the Chebyshev criterion.

There is a price to be paid for this safety, however. For example, the sum of the squares of the errors in a Chebyshev approximation will be higher than if the least-squares approximation were used.

The Chebyshev approximation tends to have its extreme errors of alternate sign and is therefore sometimes called "equiripple approximation."

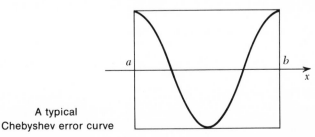

A typical
Chebyshev error curve

We begin the topic of Chebyshev approximation by the development of some theory.

13.2   Definition of the Chebyshev polynomials

The Chebyshev polynomials are defined by

$$T_o(x) = 1$$

$$T_k(x) = \cos\,(k\,\arccos\,x) \qquad (k = 1,\,2,\,\ldots\,)$$

They are intimately connected with the Fourier cosine expansion; indeed, they are the cosine function in the disguise of

$$x = \cos\,\theta$$

or

$$\theta = \arccos\,x$$

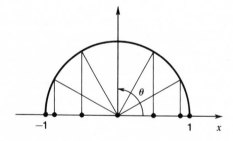

Note that if you want to define the polynomials for negative k values, then

$$T_{-k}(x) = T_k(x)$$

The $T_k(x)$ are polynomials since from elementary trigonometry

$$\cos\,n\theta = \text{polynomial of degree } n \text{ in } \cos\,\theta$$

and since

$$\cos\,(\arccos\,x) = x$$

it follows that

$$\cos{(n \arccos x)} = T_n(x)$$

is indeed a polynomial of degree n in x.
  If in the trigonometric identity

$$\cos{(n+1)\theta} + \cos{(n-1)\theta} = 2\cos{\theta}\cos{n\theta}$$

we set $\theta = \arccos x$, then we get the three-term recurrence relation for the Chebyshev polynomials,

$$T_{n+1}(x) - 2xT_n(x) + T_{n-1}(x) = 0$$

Note that the three-term relation has coefficients which are independent of n.

Conversely, if a three-term relation is independent of n, then by a translation in x plus scale factors we have the Chebyshev polynomials.

This is a useful way to the generate successive polynomials. From $T_0 = 1$, $T_1 = x$, we get, in turn,

$$\begin{aligned}
T_2 &= 2xT_1 - T_0 = 2x^2 - 1 \\
T_3 &= 2xT_2 - T_1 = 2x(2x^2 - 1) - x = 4x^3 - 3x \\
T_4 &= 2xT_3 - T_2 = 2x(4x^3 - 3x) - (2x^2 - 1) \\
&= 8x^4 - 8x^2 + 1
\end{aligned}$$

NOTES

1 *The $T_k(x)$ are alternately odd and even polynomials.*

2 *The leading coefficient is 2^{k-1}.*

  We next examine the orthogonality properties. We know from (Sec. 12.1)

$$\int_0^\pi \cos{m\theta}\cos{n\theta}\, d\theta = \begin{cases} 0 & (m \neq n) \\ \dfrac{\pi}{2} & (m = n \neq 0) \\ \pi & (m = n = 0) \end{cases}$$

Making the standard substitution to get to the Chebyshev polynomials,

$$\theta = \arccos x$$

we get

$$\int_{-1}^{1} \frac{T_m(x)\,T_n(x)}{\sqrt{1-x^2}}\,dx = \begin{cases} 0 & (m \neq n) \\ \dfrac{\pi}{2} & (m = n \neq 0) \\ \pi & (m = n = 0) \end{cases}$$

Thus, the Chebyshev polynomials form an orthogonal set over the interval $-1 \leq x \leq 1$ with weight function

$$w(x) = \frac{1}{\sqrt{1-x^2}}$$

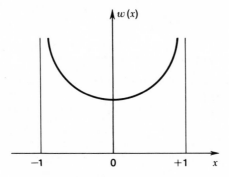

The Legendre polynomials $P_n(x)$ were orthogonal over the same interval $-1 \leq x \leq 1$ but had $w(x) \equiv 1$.

The Chebyshev polynomials are also orthogonal over the discrete set

$$x_p = \cos \frac{\pi p}{N} \qquad p = 0, 1, \ldots, N-1$$

$$\sum_{p=0}^{N-1} T_m(x_p)\,T_n(x_p) = \begin{cases} 0 & (m \neq n) \\ \dfrac{N}{2} & (m = n \neq 0) \\ N & (m = n = 0, N) \end{cases}$$

or over the same set plus one end value $p = N$ **provided** we weight the two end terms by $\frac{1}{2}$,

$$\tfrac{1}{2} T_m(-1)\,T_n(-1) + \sum_{p=1}^{N-1} T_m(x_p)\,T_n(x_p)$$

$$+ \tfrac{1}{2} T_m(1)\,T_n(1) = \begin{cases} 0 & (m \neq n) \\ \dfrac{N}{2} & (m = n \neq 0) \\ N & (m = n = 0, N) \end{cases}$$

We can also use the set of points (see Secs. 12.3 and 12.6)

$$t_p = \cos \frac{\pi}{N}(p + \tfrac{1}{2}) \qquad (p = 0, \ldots, N - 1)$$

which since they are symmetrical, seem (and can be shown) to be somewhat better.

PROBLEM 13.2

1 Discuss the point that the coefficients of the expansion are given by integrals in the continuous case and that the discrete case can be viewed as a numerical approximation to the integrals. Note that we have both the trapezoid and midpoint rules available.
2 Derive the orthogonality over the midpoint set.

13.3 The Chebyshev principle

Chebyshev proved that of all polynomials of degree n having the leading coefficient equal to 1, the Chebyshev polynomial divided by 2^{n-1}, has the least extreme value in the interval $-1 \leq x \leq 1$. In other words, no other polynomial of degree n with leading coefficient 1 will have a smaller extreme value than

$$\frac{T_n(x)}{2^{n-1}}$$

in the interval $(-1 \leq x \leq 1)$.

To prove this statement, consider the fact that $T_n(x)$ is a cosine in disguise, and in the interval $0 \leq \theta \leq \pi$ there are $n + 1$ extreme values alternating $+1$ and -1. Thus,

$$\frac{T_n(x)}{2^{n-1}}$$

also has $n + 1$ extreme values. Now, if there were a polynomial $\phi(x)$ of degree n with leading coefficient 1 which had a smaller extreme value in the interval then at the extremes of $T_n(x)/2^{n-1}$, the expression

$$\frac{T_n(x)}{2^{n-1}} - \phi(x) = \text{polynomial of degree } (n - 1) \text{ in } x$$

will be alternately (+) and (−) at the

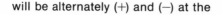

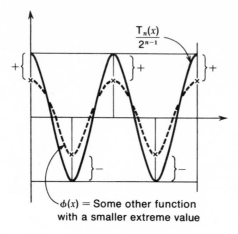

$\phi(x)$ = Some other function
with a smaller extreme value

alternate extremes and hence will have n zeros. But, as indicated, the difference is a polynomial of degree $(n-1)$ that can have n zeros only if it is identically zero; hence, the Chebyshev polynomial is the one with the minimum maximum deviation from the x axis.

This has given rise to a general concept in approximation theory, the idea of approximating a function by a sum of members of a given class (the class of polynomials in this particular case) such that the maximum error of the approximation is a mimium. The minimum maximum error is often called the "minimax."

For an expansion in Chebyshev polynomials,

$$f(x) = \sum_{k=0}^{N} a_k T_k(x)$$

$$\begin{cases} a_k = \dfrac{2}{\pi} \displaystyle\int_{-1}^{1} \dfrac{f(x)\,T_k(x)}{\sqrt{1-x^2}}\,dx & (k \geq 1) \\[4mm] a_0 = \dfrac{1}{\pi} \displaystyle\int_{-1}^{1} \dfrac{f(x)}{\sqrt{1-x^2}}\,dx & k = 0 \end{cases}$$

when we cut off the series at some point N, then provided the convergence of the expansion is rapid (a point we shall show is usually true), the error is almost exactly the first term neglected, which is a Chebyshev (equiripple) polynomial having a minimum maximum deviation from zero; hence, the complete error is almost a minimax.

The fact that the Chebyshev polynomials give, at the same time, both the weighted least-squares fit and almost the minimax fit appears to be a bit of a contradiction until it is noted that in least-squares fitting of polynomials, the largest errors are usually at the ends of the interval and that is exactly where the Chebyshev weight function puts most emphasis. For the finite sums (as opposed to the integrals) it is the unequal spacing (the x_i) that produces the effect of the weights.

PROBLEM   13.3

1 Show that

$$\left| \frac{T_k(x)}{2^{k-1}} \right| \le \frac{1}{2^{k-1}} \qquad (0 \le x \le 1)$$

has x^k as leading term and for k sufficiently large this polynomial is essentially zero on the whole interval, and that this suggests that the powers of x are not linearly dependent practice.

13.4   Various identities

Since the Chebyshev polynomials are both orthogonal polynomials and cosines in disguise, they naturally satisfy many different identities.
  For example, the trigonometric identity

$$\cos (m + n)\theta + \cos (m - n)\theta = 2 \cos m\theta \cos n\theta$$

is equivalent to

$$T_{m+n}(x) + T_{m-n}(x) = 2T_m(x)T_n(x)$$

Another example

From
$$\cos 2x = 2 \cos^2 x - 1$$

we have
$$\cos 2mx = 2 \cos^2 mx - 1$$

or
$$T_{2m} = 2T_m^2 - 1$$

which is a special case of
$$T_k \{T_m(x)\} = T_{km}(x)$$

Again from differentiating

$$T_{n+1}(x) = \cos\left[(n+1)\arccos x\right]$$

we get

$$\frac{1}{n+1}\frac{dT_{n+1}(x)}{dx} = -\sin\left[(n+1)\arccos x\right]\frac{-1}{\sqrt{1-x^2}}$$

and

$$\frac{1}{n-1}\frac{dT_{n-1}(x)}{dx} = -\sin\left[(n-1)\arccos x\right]\frac{-1}{\sqrt{1-x^2}}$$

Subtracting the lower from the upper

$$\frac{1}{n+1}T_{n+1} - \frac{1}{n-1}T_{n-1} = \frac{1}{\sqrt{1-x^2}}\left[\sin(n+1)\theta - \sin(n-1)\theta\right]$$

$$= \frac{2\cos n\theta \sin\theta}{\sin\theta}$$

$$= 2T_n(x) \qquad (n>1)$$

From this we can find the indefinite integrals

$$\int T_0(x)\,dx = T_1(x) + C$$

$$\int T_1(x)\,dx = \tfrac{1}{4}T_2(x) + C$$

$$\int T_k(x)\,dx = \tfrac{1}{2}\left[\frac{T_{n+1}(x)}{n+1} - \frac{T_{n-1}(x)}{n-1}\right] + C \qquad (n>1)$$

It is almost true that any desired identity can be found by considering the proper corresponding trigonometric identity.

PROBLEMS   13.4

Show that:

1 $T_N'(x) = \displaystyle\sum_{k=0}^{N} T_{-N+2k-1}(x)$

2 $2^k x^k T_n(x) = T_{n+k} + C(k,1)T_{n+k-2} + C(k,2)T_{n+k-4} + \cdots + T_{n-k}$

3 $\dfrac{T_{2k+1}(x)}{x} = 2T_{2k} - 2T_{2k-2} + \cdots \pm T_0$

4 $T_k\{T_m(x)\} = T_{km}(x)$

13.5 Direct method of Chebyshev approximation

We can find a Chebyshev approximation (fairly accurately) if we can find the coefficients in the formal expansion. These are given by

$$a_k = \frac{2}{N} \sum_{p=0}^{N-1} F(x_p) T_k(x_p) \qquad (k = 1, \ldots, N-1)$$

$$a_0 = \frac{1}{N} \sum_{p=0}^{N-1} F(x_p) T_0(x_p) = \frac{1}{N} \sum^{1} F(x_p)$$

where the points t_p are now labeled x_p,

$$x_p = \cos \frac{\pi}{N} (p + \tfrac{1}{2})$$

provided we use

$$F(x) = \sum_{k=0}^{N-1} a_k T_k(x)$$

This provides a powerful method for finding Chebyshev expansions. Note that the $F(x)$ values need be found only once, and we can then get the coefficients a_k by the fast Fourier transform method.

PROBLEM 13.5

1 Give the formula for the other approximation, namely, equally spaced points.

13.6 Another approach

Often, a function is given in its power series representation. Suppose we take enough terms in the power series to be well inside the necessary accuracy, say $M + 1$ terms,

$$F(x) = \sum_{k=0}^{M} a_k x^k$$

To convert this to a Chebyshev series, we first write it as

$$a_0 + x \{a_1 + x [a_2 + \cdots + x(a_{M-1} + a_M x) + \cdots]\}$$

Starting in the inner parentheses, we have

$$a_{M-1} + x a_M = a_{M-1} T_0 + a_M T_1(x)$$

Assuming that at the kth stage we have (the α_i depend on k)

$$\alpha_0 T_0 + \alpha_1 T_1 + \cdots + \alpha_k T_k$$

we multiply this by x and add the next a_{M-k-1} to it. By using

$$x T_0 = T_1$$
$$x T_k = \frac{T_{k+1} + T_{k-1}}{2}$$

this becomes

$$\left(a_{M-k-1} + \frac{\alpha_1}{2}\right) T_0 + \left(\alpha_0 + \frac{\alpha_2}{2}\right) T_1 + \left(\frac{\alpha_1}{2} + \frac{\alpha_3}{2}\right) T_2$$

$$+ \cdots + \frac{\alpha_{k-1}}{2} T_k + \frac{\alpha_k}{2} T_{k+1}$$

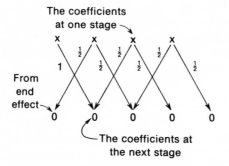

Thus we go step by step until finally we have Chebyshev expansion. Note that in the process the original $a_M x^M$ gives rise to

$$\frac{a_M}{2^{M-1}} T_M(x)$$

It is because of this effect that one says "generally speaking the Chebyshev expansion converges very rapidly."

We now examine the coefficients and drop all those which are too small to matter, using the obvious fact that

$$|T_k(x)| \le 1 \qquad \text{for all } x \qquad (-1 \le x \le 1)$$

This process is called "economization."

We can then, if we wish, reconvert to a power series and thus find a very low degree polynomial which does about as good a job as the much higher-order truncated power series did—a great saving in machine time for a function that is going to be computed many, many times. This process (or more elaborate ones) is usually adopted for the most commonly computed special functions.

13.7   The leveling process

Often, the Chebyshev approximation found by the above methods is satisfactory in spite of the fact that the error curve is not exactly equiripple (because the error is not just the first term neglected but is, in fact, the sum of all the rest of the terms of the infinite expansion). Occasionally, the result is "adjusted" to produce an exact equiripple error curve.

One approach is related to our optimization processes. We find the extremes of the error curve and equate these with an unknown linear sum of the errors generated by small changes in the coefficients of the present expansion. To do this, we essentially perturb each coefficient and note the corresponding change in the error at the current extreme values. We then ask for the sum that will exactly erase the inequalities.

The second method is to try moving the zeros of the error curve to produce an equiripple error. Both methods are iterative, and in both, the final error size is not known in advance.

RANDOM PROCESSES

14.1 Introduction

The idea of using randomness in a creative, useful way comes as a surprise to many people. Past training has usually been to try to avoid, or at least minimize, the effects of randomness in doing experiments. One purpose of this chapter is to illustrate a few of the many useful ways of handling random processes.

Example

The simple tossing of a coin and calling "heads" "1" and "tails" "0" leads to an expected value of $\frac{1}{2}$—which is neither value! This leads to the idea of a sequence of trials, and this to an infinite sequence.

But to each infinite sequence there is a binary number having the same digits in the same order. Thus, instead of considering a single number, we are led to the (noncountable) set of all binary numbers x, $0 \le x < 1$.

The concept of "random" is basically an intuitive one. The mathematician's idea of a "random process," or more accurately a "stochastic process," is an ensemble of functions

$$f_\lambda(t)$$

where the functions are indexed by the variable λ (which is often over a nondenumerable set) having a probability distribution, and t is regarded as time. As we emphasized in Chap. 1, the computing machine is finite both in size and in speed, and when we try to represent the mathematician's infinitely long numbers on a machine, we are forced to approximate them with a finite length and thus commit a roundoff error. Similarly, when we try to represent some of the infinite processes of the calculus, such as integration, differentiation, continuity, infinite series, and so forth, we are forced by the finite speed of the machine to make a truncation error. This time, faced with a stochastic process which is an infinite process, we are forced to approximate it with some finite number of steps and use only a finite representation at each step.

There are conceptually at least two stages of descent from the stochastic process to what we shall use to simulate it. The first step consists in selecting one member of the ensemble of random functions; the second step consists in selecting a finite sequence from this infinitely long function of time. As always, our representation of t and the values of $f_\lambda(t)$ are also finite, and often this is a third stage in the descent. Thus what we shall do on the machine is rather remote from what the mathematical idea represents. If the theory were more adequately developed, we should have a corresponding "stochastic error," "sampling error," or some such term to measure the error made by the replacement.

The confusion that arises from this reduction from the infinite to the finite is widespread. Thus, a reviewer of the Rand table of "One Million Random Digits and One Hundred Thousand Normal Deviates"† suddenly realized, in midstream as it were, that now that the table was published, the numbers were perfectly predictable and hence theoretically could not be random!

Let us turn from the grand picture of a stochastic process to the simpler idea of a sequence of random digits. The intuitive idea of a sequence of random digits is that there is no discernible pattern in the sequence, that the next integer is not predictable from the past ones.

$$3, 7, 9, 2, 4, \ldots$$

This is an essentially negative idea and hence cannot be proved to be true for any specific finite sequence. All we can do in practice is to check that some of the more obvious patterns do not occur.

Without explicitly saying it, we have implied that the various digits, say 0, 1, . . . , 9, occur with equal frequency—well, not quite **equal** since if we had a sequence of 100 random digits, we should be somewhat surprised if there were exactly 10 of each. How unequally frequent should we allow them to be? We are not quite sure; there are some statistical theories to indicate what fluctuations we might expect and how often they might occur, but about all we **know** is that we do not want a very unlikely sample, and this is a very weak remark indeed. For example, **the sequence**

$$9, 9, 9, 9, 9, \ldots, 9$$

(100 in all) **is just as likely as any other specific sequence**, say

$$3, 7, 4, 2, 9, \ldots, 7$$

---

†The Free Press of Glencoe, Ill. Chicago, 1955.

but we should probably reject the first and we might quite likely accept the second if it were presented to us for use as a finite part of an infinite sequence of random digits. And our intuition is not wrong in this matter; regardless of all the theory that may be presented to the contrary, basically a finite sequence of numbers is random only with respect to the use that is going to be made of them.

The practical use of randomness requires a fine balance of "common" and "uncommon" sense, but since we do not want to examine every use of random numbers carefully, it is convenient to have some well-tested sequence available in a library routine. In a very important application the common sense **must** be applied by some knowledgeable person.

14.2 Buffon's needle

Count Buffon in 1773 observed that if a needle of length l were "thrown down at random" on a flat, horizontal surface ruled with lines at unit spacing, then the mathematical probability of a needle's crossing a line would be $2l/\pi$. From this he reasoned he could, by repeated random trials, "experimentally determine the value of π."

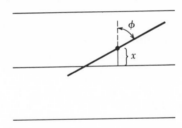

Let us first derive this formula. Suppose that the family of parallel, equally spaced, straight lines have a spacing of unit length. Let the needle have length $l \le 1$. When we throw it down at random, the center of the needle can be at any distance from 0 to $\frac{1}{2}$ from some line. Let this be the variable x. Next consider the angle ϕ at which the needle lies. We consider the variables x and ϕ random, independent, and uniformly distributed in their ranges; that is what we meant by "thrown down at random."

If we are to have an intersection of the needle with the line, then

$$x \le \frac{l}{2}\cos\phi \qquad -\frac{\pi}{2} < \phi < \frac{\pi}{2}$$

The shaded area in the figure is the area in which x and ϕ must lie if there is to be an intersection.

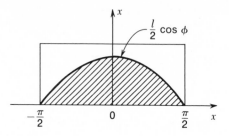

The ratio of this area to the total area is the probability p of a crossing. This ratio is

$$p = \frac{\displaystyle\int_{-\pi/2}^{\pi/2} (l/2) \cos \phi \ d\phi}{\frac{1}{2}\pi} = \frac{l}{\pi} \int_{-\pi/2}^{\pi/2} \cos \phi \ d\phi = \frac{2l}{\pi}$$

14.3   Monte Carlo methods

Buffon's observation is an example of what is called a **Monte Carlo method** of using a random process (throwing the needle at random) to determine a nonrandom result (in this case π). Now, if the value of π were only known to be between 1 and 10, this would be an excellent way to find that π is around 3. With some labor and many trials, we might get a value around 3.1. But to try to get greater accuracy would tax our ability to draw the lines accurately, to have the needle just the right length, to throw the needle at random, to count the questionable crossings, and so forth. And this is a general observation. Monte Carlo methods are perhaps most effective in the preliminary stages, where they help to give a general idea of a situation, but they are of less value if accurate results are desired.

Evidently, if we had a source of random numbers in a computing machine, we could simulate Buffon's experiment and compute the value of π. The accuracy would, of course, depend on the sampling fluctuations and on our source of random numbers. Indeed, it would in the long run turn out to be more of a test of the random number generator than it would be a measure of the value of π.

Monte Carlo methods have produced a vast literature, but the author observes that in practice their importance has been greatly exaggerated. The random numbers are much more often used to simulate random (stochastic) processes, especially noise, which will be examined later in Sec. 14.8.

14.4 Generation of random numbers

What can we mean by "a random number between 0 and 1"? In the mathematical sense if you did have a random number between 0 and 1, you could never tell anyone which number it was because you could never completely describe it.

A "random number"

$$0.1736497 \cdots$$

has an infinity of digits and no short description can serve to describe it.

In a practical sense what we should probably mean is that in, say, the fixed-point notation of a k-digit binary computing machine (which has 2^k possible numbers) we shall select one number at random (a rather circular definition). Actually, we shall in the end not even do this well; rather, from the set

$$1, 5, 9, 13, \ldots, (2^k - 3)$$

or from the set

$$3, 7, 11, 15, \ldots, (2^k - 1)$$

we shall select the numbers in some sequence until we exhaust the entire set of 2^{k-2} numbers, and then we shall repeat the sequence **exactly**. For a 35-digit machine this period of repetition is

$$2^{33} \sim 10^{10} = 10 \text{ billion}$$

so we need not get too excited over the repetition (though it is necessary to consider this in some extreme cases). It is the lack of possible repetition in the sequence that should bother us more; we are, as it were, sampling in the given set **without** replacement.

Our method for generating the random numbers will be quite simple. We shall choose a starting value x_0 from the set and compute at the ith step the next random number

$$x_{i+1} = \rho x_i \qquad \text{modulo } 2^k$$

which means, "keep the last k digits of the $2k$ (or $2k - 1$) digit product ρx_i." This is simple to do on binary computers. As we shall show, the value of ρ must be of the form

$$\rho = 8t \pm 3 \qquad \text{for some } t$$

if we are to get a maximally long sequence before repeating. A good choice for ρ is around

$$01 \cdots 1$$

or

$$10 \cdots 1$$

and it should probably have about half 0s and half 1s in it.
    We shall also later discuss the proper choice of x_0.

Example   *For a $k = 5$ binary digit machine*

Pick

$$\rho = 10101$$
$$x_0 = 10001$$
$$\rho x_0 = 101100101$$
$$\rho x_0 \,(\text{mod } 2^5) = 00101$$

binary	decimal
$x_0 = 10001$	17
$x_1 = 00101$	5
$x_2 = 01001$	9
$x_3 = 11101$	29
$x_4 = 00001$	1
$x_5 = 10101$	21
$x_6 = 11001$	25
$x_7 = 01101$	13
$x_8 = 10001$	17

14.5   Proof that the generator works

The standard notation in number theory

$$x \equiv a(\text{mod } m)$$

means that $x - a$ is divisible by m.
    If we have a k-digit binary machine and take the last k digits of the double-length product of ρx_n, then we want

$$x_{n+1} \equiv \rho x_n (\text{mod } 2^k) \qquad (k \geq 3) \tag{14.1}$$

Now if 2 divides x_0 (the starting value), then the sequence will be equivalent to

$$y_0 = \frac{x_0}{2}$$

$$y_{n+1} \equiv \rho y_n (\text{mod } 2^{k-1})$$

which is equivalent to a shorter word length, and we are wasting machine capacity. Hence, we take x_0 to be an odd number.

Similarly, ρ should be odd, since if ρ were even, then

$$x_{k+1} = \rho^{k+1} x_0 = 0$$

and we should have a trivial sequence of all zeros past this point.

Now any odd number ρ can be written in one of the forms

$$8t - 3 \qquad 8t - 1 \qquad 8t + 1 \qquad 8t + 3$$

for some t.

Theorem 1 *If $\rho = 8t \pm 1$, then*

$$\rho^{2^{k-3}} \equiv 1 (\text{mod } 2^k) \tag{14.2}$$

*(that is, the **order** of ρ is a divisor of 2^{k-3}).*

Proof This is equivalent to saying

$$2^k \text{ divides } \rho^{2^{k-3}} - 1$$

Since, in general,

$$a^2 - 1 = (a + 1)(a - 1)$$

then by repeated use of this identity

$$\rho^{2^{k-3}} - 1 = (\rho^{2^{k-4}} + 1)(\rho^{2^{k-5}} + 1) \cdots (\rho + 1)(\rho - 1) \tag{14.3}$$

For each term $(i \geq 1)$

$$\rho^{2^i} + 1 = (8t \pm 1)^{2^i} + 1$$

$$= 1 + (1 \pm 8t)^{2^i}$$

$$= (1 + 1) + \sum_{k=1}^{2^i} (\pm 1)^k C(2^i, k) t^k 8^k$$

where $C(\alpha,\beta)$ is a binomial coefficient. From this it follows that 2 divides $\rho^{2^i} + 1$ and leaves an odd number, and so no higher power of 2 divides it. From (3),

$$2^{k-4} \text{ divides } (\rho^{2^{k-4}} + 1)(\rho^{2^{k-5}} + 1) \cdots (\rho^2 + 1) \qquad (14.4)$$

which leaves the factors $(\rho + 1)(\rho - 1)$. We have

$$\begin{aligned}
(\rho + 1)(\rho - 1) &= \rho^2 - 1 \\
&= (8t \pm 1)^2 - 1 \\
&= 16(4t^2 \pm t)
\end{aligned}$$

Hence,

$$2^4 \text{ divides } (\rho + 1)(\rho - 1)$$

Now, using this with (14.4) in (14.3), we have

$$2^4 \times 2^{k-4} = 2^k \text{ divides } \rho^{2^{k-3}} - 1$$

Theorem 2   *If $\rho = 8t \pm 3$, then*

$$\rho^{2^{k-3}} \not\equiv 1 \,(\text{mod } 2^k)$$

$$\rho^{2^{k-2}} \equiv 1 \,(\text{mod } 2^k)$$

(that is, the order of ρ is exactly 2^{k-2}).

Proof   Again using (14.3), we have

$$\rho^{2^i} + 1 = 1 + (3 \pm 8t)^{2^i}$$

$$= 1 + 3^{2^i} + \sum_{k=1}^{2^i} C(2^i,k) t^k (\pm 8t)^k 3^{2^{i-k}}$$

But

$$\begin{aligned}
1 + 3^{2^i} &= 1 + (4 \overset{\scriptstyle \angle}{-} 1)^{2^i} \\
&= 1 + (1 - 4)^{2^i}
\end{aligned}$$

$$= 1 + 1 + \sum_{k=1}^{2^i} C(2^i,k)(-4)^k$$

Hence,

$$\rho^{2^i} + 1 = 2 + \sum_{k=1}^{2^i} C(2^i,k)\left[3^{2^{i-k}}(\pm 2t)^k + (-1)^k\right]4^k$$

Thus,

$$2 \text{ divides } \rho^{2^i} + 1$$

but

$$4 \text{ does not divide } \rho^{2^i} + 1$$

We therefore have from (14.3)

$$2^{k-4} \text{ divides } (\rho^{2^{k-4}} + 1)(\rho^{2^{k-5}} + 1) \cdots (\rho^2 + 1)$$

but 2^{k-3} does not divide it.

We have still to consider the terms $(\rho + 1)(\rho - 1)$. Observe that when $\rho = 8t \pm 3$, $(\rho + 1)(\rho - 1) = 8(8t^2 \pm 6t + 1) = 8$, which is an odd number.

Thus, in either case,

$$2^3 \text{ divides } (\rho + 1)(\rho - 1)$$

and 2^4 does not. Therefore, combining the two results,

$$2^{k-1} \text{ divides } \rho^{2^{k-3}} - 1$$

but 2^k does not. In other words,

$$\rho^{2^{k-3}} \not\equiv 1 \,(\text{mod } 2^k)$$

We now write

$$\rho^{2^{k-2}} - 1 = (\rho^{2^{k-3}} + 1)(\rho^{2^{k-3}} - 1)$$

and

$$2 \text{ divides } \rho^{2^{k-3}} + 1$$

$$2^{k-1} \text{ divides } \rho^{2^{k-3}} - 1$$

so that

$$2^k \text{ divides their product}$$

and

$$\rho^{2^{k-3}} \equiv 1 \,(\text{mod } 2^k)$$

Theorem 3   *If $\rho = 8t - 3$, then the sequence*

$$x_0, x_1, x_2, \ldots, x_{2^{k-2}-1}$$

generated by (1) *is some permutation of*

$$1, 5, 9, \ldots, (2^k - 3) \qquad \text{if } x_0 \equiv 1 \,(\text{mod } 4)$$

or

$$3, 7, 11, \ldots, (2^k - 1) \qquad \text{if } x_0 \equiv 3 \,(\text{mod } 4)$$

Proof   Consider the values $(x_0 \rho^n)$, $n = 0, 1, \ldots, 2^{2^{k-2}} - 1$. The difference between consecutive terms

$$
\begin{aligned}
x_0 \rho^{n+1} - x_0 \rho^n &= x_0 \rho^n (\rho - 1) \\
&= x_0 \rho^n (8t - 4)
\end{aligned}
$$

is divisible by 4. But we know (theorem 2) that the order of ρ is 2^{k-2}; hence, we have 2^{k-2} distinct terms whose differences are divisible by 4, and the theorem follows.

Thus, we see that, depending on how we select x_0, provided that we pick $\rho = 8t - 3$ (for any value t), we can get a permutation of one of the sequences in theorem 3.

Although we have regarded the x_n as integers in the derivation, when we come to use them, we can, if we please, place the binary point on the extreme left; effectively, we use

$$x_n . 2^{-k}$$

But it is not satisfactory to select just any value for t when selecting the ρ. For example, $t = 1$ gives $\rho = 5$, and whenever a small value of x_n occurs, say $1 \cdot 10^{-k}$, then there will be a long sequence of gradually increasing numbers following it. This suggests that probably we want to pick a t so that ρ has leading digits either $01 \ldots$ or $10 \ldots$ to avoid these troubles. There have been a number of careful studies of the advantages of choosing different permutations of the same numbers, but the question has apparently not been difinitively settled as yet (1970).

For a given $\rho = 8t - 3$, we have two cycles, each of which exhausts one-fourth of the available numbers. The remainder of the available numbers occur in smaller cycles.

If, in the sequences of theorem 3, the last two binary digits are dropped, then the resulting sequence is a permutation of 0, 1, 2, . . . , $2^{k-2} - 1$ having a full cycle of length 2^{k-2}. If we do not drop the last two digits, then the last digit is always a 1 and certainly is not random. As we progress from right to left and examine the various trailing digits, we find less and less structure. Thus, it is customary not to depend on the last few digits' being random.

The testing of a random-number generator is an arcane art and we will not discuss it in detail; some obvious properties are the appropriate frequency of each assigned pattern of digits, a correlation coefficient of almost zero for successive digits and in digits lagged by k, that is,

$$E\{(x_{i+k} - \tfrac{1}{2})(x_i - \tfrac{1}{2})\} = 0$$

and also a "flat power spectrum."

You can wear out a large computing machine doing all the tests you can think of after a few days of meditating.

14.6 Generation of random numbers on a decimal machine

The basic results in this area are in Moshman's paper,[†] which gives the formula for an s-digit decimal machine ($s > 4$)

$$\rho = 7^{4k+1}$$

which has a period[‡] of length $5 \cdot 10^{s-3}$, and also the results of testing it on an 11-digit machine. Another choice[§] of ρ,

$$\rho = 76, 768, 779, 754, 638, 671, 877 \qquad \text{reduced mod 10s}$$

gives maximal periods.

[†]J. Moshman, The Generation of Pseudo-Random Numbers on a Decimal Machine, *J. Assoc. Computing Machinery*, vol. 1, pp. 88–91, 1954.

[‡]E. Bofinger and V. J. Bofinger, On a Periodic Property of Pseudo-Random Numbers, *J. Assoc. Computing Machinery*, vol. 5, pp. 261–265, 1958, amend Moshman's results slightly.

[§]From E. N. Gilbert.

Other investigations with the use of more elaborate formulas have been made. For a binary machine,

$$x_{n+1} = (2^a + 1)x_n + C(\text{mod } 2^{35})$$

can be used.

Also studied are formulas of the form†

$$x_{n+1} = \alpha x_n + \beta x_{n-k}$$

The problem is evidently in a great state of flux and cannot be settled here at this time. Indeed, there are reasons to believe that these methods can never be completely satisfactory.‡

14.7   Other distributions

Sometimes we want random numbers selected from other distributions than the equiprobable, flat distribution that we generate by our random-number generator.

The most commonly desired distribution is a "normal distribution density"

$$\frac{2}{\sqrt{\pi}} e^{-x^2/2}$$

One easy way to obtain this is to add 12 numbers from the flat distribution and then subtract 6 from the sum.

$$y = x_1 + x_2 + \cdots + x_{12} - 6$$

The distribution density of y is close to normal, having mean zero and a variance of 1. The tails reach from -6 to $+6$, thus "out to 6σ," and beyond that there are exactly zero occurrences.

Flat distribution

$$Mean = \mu = \frac{\displaystyle\int_0^1 x\, dx}{\displaystyle\int_0^1 dx} = \frac{1}{2}$$

---

†See D. E. Knuth, "Seminumerical Algorithms," Addison-Wesley Publishing Company, Inc., Reading, Mass., 1969, for an exhaustive account of random-number generation.

---

‡Marsaglia has shown that all congruential generators have a certain flaw, but its relevance in most applications is not clear.

Variance

$$\sigma_i^2 = \frac{\displaystyle\int_0^1 (x - \tfrac{1}{2})^2 dx}{\displaystyle\int_0^1 dx}$$

$$= \frac{(x - \tfrac{1}{2})^3}{3}\Big|_0^1$$

$$= \frac{1}{12}$$

From the independence of the x_i

variance of sum = sum of variances

$$\sigma^2 = 12 \times \tfrac{1}{12} = 1$$

Probability of exceeding 6σ is approximately 2×10^{-9}.

This means that the extremely unlikely events cannot occur in our model, though they can theoretically occur in the mathematical distribution. This difference usually favors our practical distribution rather than the theoretical one.

Suppose that we want some other distribution $f(y)$; how can we get it from a flat distribution? One way is to equate the two cumulative distributions, the flat one in the variable x and the desired one in y,

$$\int_0^x 1 \, dx = \int_0^y f(y) \, dy = F(y) = x$$

Thus, if we can find the inverse function of $F(y)$, we have

$$y = F^{-1}F(y) = F^{-1}(x)$$

In principle, this is all that is required.

A simple example of this method is the exponential distribution

$$f(y) = e^{-y}$$

We have

$$F(y) = \int_0^y e^{-y} \, dy = 1 - e^{-y}$$

or

$$e^{-y} = 1 - x$$

In practice, we can replace the random number $1 - x_i$ with x_i so that we have $e^{-y_i} = x_i$, or

$$y_i = -\ln x_i$$

This has been used with excellent success.

Another example

$$f(y) = \cos y \qquad 0 \le y \le \frac{\pi}{2}$$

$$\int_0^{\pi/2} f(y) \, dy = \sin y \Big|_0^{\pi/2} = 1$$

Hence

$$F(y) = \sin y = x$$
$$y_i = \arcsin x_i$$

14.8 The use of random numbers to simulate noise

Probably the most common use of random numbers is to simulate the noise that arises in the course of doing a simulation of some process, radar signal, circuit, and so forth. It should be evident that we often have to shape the numbers we use to fit the model of the noise. Sometimes only simple operations are necessary, but in complicated simulations there may be some inner structure, say, some "color" in the spectrum of the noise, that must be simulated. The theory of generating such a stream of numbers, having a preassigned internal structure, lies outside of this elementary text.

In such simulations of noise an enormous number of random numbers can easily be used, and the finite length of the cycle of the generator can occasionally have a serious effect. One could go to double precision, to a pair of random-number generators (with a third used to determine which of the two to use next, provided the obvious faults of this are eliminated), in order to get a sufficiently long period before repetition.

Experience seems to indicate that the careless use of a random-number generator will trap the unwary many more times than at first seems reasonable, but also that with care random-number generators give very satisfactory results. It is a field for the beginner to be wary and suspicious of and he should make a few careful checks that all is going as he thinks it should before he plunges into the details of using the simulation as a working tool of design optimization.

Another frequent use of random numbers is to make a "random choice" in some process. Here only a few of the first bits of a generated random number are used to make the choice. Traffic studies, random sampling, and process simulation are typical examples of application.

15

15.1 Why new library routines may be needed

The material in the preceding chapters has been selected for what is believed to be most needed by the practicing engineer or scientist, and not for what is currently in textbooks and the literature. Therefore, it should not be surprising that the material is sometimes new, and not in the local library. Even when the material is well known, the presentation may be unusual.

The selection of the material was also based on the observation that the scientist or engineer has a limited amount of time and effort that he can devote to the topic of numerical methods, and that although more material would of course be useful, it is often necessary to be practical and severely limit the chosen material. Thus, there is no pretense that the material selected will meet **all** the needs that may arise, it is merely hoped that it will meet some of the more common needs.

No attempt was made to pick the "best" methods in the sense of the efficient use of the machine time and capacity; rather, the methods were selected for the efficient use of the learner's time and effort. Furthermore, no attempt was made to be complete; most of the chapters could be, and some have been, the subject of a one-volume (or more) work.

Thus, the user of this book may often find that what he wants is not in the local computer library and he must consider writing a program for the library. A few remarks of warning are therefore necessary.

15.2 Theories of library design

Most computing centers sooner or later decide to gather together a library of the "better" routines so that the average user can have the approved processes readily available and can avoid ones known to be "bad." Behind the selection of the material for the library are various beliefs, which, unfortunately, are seldom clearly stated. Among the pure principles are the following, though in practice many different principles are generally used in the selection of many particular library routines.

1. *Get Rid of the Problem*   The first method that all too often is

used is simply to write something that is supposed to work so that the problem of supplying a library routine will be finished. This approach is most often used by professional programmers with no interest in, and usually little knowledge of, numerical methods. Unfortunately, the above has too often been an accurate characterization of the programs that were issued by the manufacturers of computers.

2. *Save Machine Time*   Because the first routines received from the manufacturer were so exceedingly slow, the local group often wrote some very fast ones, but without considering the breadth of coverage that may be asked for, or the accuracy of the results. Indeed, often the answers were clearly wrong for simple problems.

3. *Mathematical Completeness*   Another theory, which comes mainly from a revulsion from the first two, is to write a program that both will cover every possible situation that can arise and will be sure to get "the right answer." This approach is likely to be taken by the pure mathematician who has recently become interested in computation. Such programs, if they work, are apt to consume large amounts of time covering the pathological cases that are more often conceptual than real (so far as practice is concerned), and large amounts of storage as well. The underlying principles of the routine are apt to be very hard to understand, and the answers, though "right," are often not appropriate. And almost always the mathematical proof that the method works ignores the actual numbers used by computers and hence to some extent is irrelevant.

4. *Do Not Offend Anyone*   This approach, instead of trying to cover every circumstance in a single library routine, supplies a wide assortment of different routines covering somewhat the same ground but with each one having special properties and advantages in special situations. The user is thus faced with a list, say, of a dozen slightly differing routines for integrating a system of ordinary differential equations, and he does not know which one to choose. This guiding principle allows the computer-center group to avoid the responsibility of selecting the better routines and thus to avoid offending anyone.

5. *Garbage Disposal*   Another method is to decide that the various programs in method 4 can be combined (without regard to the quality of the individual parts) into one program with the user's having the option to select the particular details that he wants. This can produce a very long calling sequence for the library routine. In order not to baffle the beginner, it is sometimes arranged that he only needs to mention the sequence variations that he wants; those he does not want he omits and thus gets the standard version in those parts. The describing memo is apt to be very thick and discouraging to read.

6. *Black-box Library*   There is a school of thought which believes that the user need not understand the inside workings of the

routine. But this is obviously false. In so simple and common a routine as that for finding the sine of an angle, unless the user realizes that in one way or another the input angle is converted to rotations and the integer part (in rotations) is dropped, he is not in a position to realize that for very large angles there must be a significant loss of accuracy. With more involved routines than for the sine it is even more difficult to understand what you have obtained if you do not understand the basic plan of solution. Thus, there is a real virtue in having library routines that are easy to understand; they save the user a lot of time and trouble when problems arise. Clear and easily read documentation is an important issue often overlooked by library planners.

7. *A Guaranteed Library*   Another school of thought believes that every library routine should give answers accompanied with guaranteed error bounds. The quality of the error bounds may be dubious, but they must be rigorous bounds. Again, this luxury is nice to have, but is quite likely to cost a good deal of machine speed and storage, and for the user it is apt to be rather difficult to understand what is being done and how seriously he should take the reported bounds as being indicative of the probable error. The theory also supposes that the user looks at the output of the library computation before going on and that the routine is not buried inside a very large program, where in practice the error bounds are all too often ignored by the subsequent computation.

8. *Trick Methods*   Another difficulty with some routines is that when machine optimization is made the paramount goal then the library routine is very likely to be machine dependent, not only in particular details, but often in the basic plan of attack. The user suffers in such circumstances because he cannot hope to know why the program is written the way it is.

9. *Answer the Right Question*   Still another method, and one that the author of this book obviously believes in, is to try to have the library routine answer the **probably** asked question, not some apparently equivalent mathematical question. The finding of "multiple roots as multiple ones and not close distinct ones" is perhaps the best example of trying to find the right question that the library routine is supposed to answer rather than the question that is so often asked.

10. *A Library Is a Library*   The author also believes that a library of routines is **not** a collection of isolated programs, but rather must have an overall plan plus *both* a common internal structure and a common external structure in its programs. Thus, there should be a common level of "error returns," uniform protection against errors, error bounds that mean the same thing when they occur in various routines, and a uniform "calling sequence" structure as well. When possible, the same basic mathematical tools should be used (even at the cost of some machine efficiency) to reduce the burden of learning that falls on the user.

With all the criteria for selecting routines for a library, it should be clear that we are still very far from having any agreed-upon basis for selection. Unfortunately, we do not even know enough about what we want to do in order to say what cases we want to handle and with what frequencies the various subcases occur; hence optimization cannot be seriously considered. As an example, how often will multiple roots occur? If seldom, then a method that is very fast on simple zeros and rather slow on multiple ones may be preferable to one that is a bit slower on simple zeros but much faster on multiple ones. And so it goes; we simply do not know how to balance the gain in one characteristic against the loss in another, let alone what characteristics we most want. Too often machine speed is made the goal, rather than user efficiency.

Meanwhile, something must be done and various groups do what they can to establish the local library routines. Little is being done to approach any general processes for selection of library routines. Often the selectors themselves have only a slight awareness of the bases for their preferences. Thus, the user must still wonder on what bases his local library routine was selected. All that can be said at this point is "beware."

index